INTRODUCTION TO CIRCUITS AND SYSTEMS

Edited by
Aldon Cris B. Galido

Delve Publishing

Introduction to Circuits and Systems

Edited by Aldon Cris B. Galido

Edition 2018

Delve Publishing

Published by
Delve Publishing LLC
708 3rd Avenue, 6th Floor
New York
NY 10801
United States of America
www.delvepublishing.com

ISBN: 978-1-68095-903-1 (Hardback)

Printed in United States of America

Library of Congress Control Number: 2016920462

About the Editor

Aldon Cris B. Galido

Aldon is currently taking up his MS EE (Microelectronics) degree at the University of the Philippines - Diliman Campus. His interests include low voltage and low power integrated circuits for wireless sensor node applications. He is currently working as a Research Fellow for the Resilient Sensory Swarms for Smart Energy and Environmental Monitoring (RESE2NSE) project at the Philippine-California Advanced Research Institute.

Table of Contents

2 TRANSISTORS

3 CAPACITORS

4 MOSFET AMPLIFIERS

5 ELECTRONIC OSCILLATOR

6 INTEGRATED CIRCUITS

7 TELECOMMUNICATION SYSTEMS

8 INSTRUMENTATION SYSTEM

9 MICROELECTROMECHANICAL SYSTEMS

List of Abbreviations

AC	Alternating Current
ADC	Analog to Digital Converters
ATE	Automated Test Equipment
BGA	Ball Grid Array
BJT	Bipolar Junction Transistor
CMOS	Complementary Metal Oxide Semiconductor
CNFET	Carbon Nanotube Field Effect Transistor
DC	Direct Current
ECAD	Electronic Computer Aided Design
ECE	Electrochemical Etching
EDA	Electronic Design Automation
EOSFET	Electrolyte Oxide Semiconductor Field Effect Transistor
ESD	Electrostatic Discharge
ESR	Equivalent Series Resistance
FDM	Frequency Division Multiplexing
FET	Field Effect Transistor
HEMT	High Electron Mobility Transistor
HVDC	High Voltage Direct Current
IC	Integrated Circuit
IFSET	Ion Sensitive Field Effect Transistor
IGBT	Insulated Gate Bipolar Transistors
ITRS	International Technology Roadmap for Semiconductors
ITU	International Telecommunication Union

JFET	Junction Gate Field Effect Transistor
LED	Light Emitting Diodes
LFO	Low Frequency Oscillator
LGA	Land Grid Array
MESFET	Metal Semiconductor Field Effect Transistor
MOS	Metal Oxide Semiconductor
NEMS	Nano-Electromechanical Systems
OFET	Organic Field Effect Transistor
PQFP	Plastic Quad Flat Pack
RF	Radio Frequency
RFI	Radio Frequency Interference
RFID	Radio Frequency Identification
RIE	Reactive Ion Etching
SAW	Surface Acoustic Wave
SEM	Scanning Electron Microscope
SMD	Surface Mount Device
TSOP	Thin Small Outline Package
VCO	Voltage Controlled Oscillator
VCO	Voltage Controlled Oscillators
WSI	Wafer Scale Integration

Preface

The electronic circuits and systems program involves the study of the processes of analysis and design of electronic circuits and systems. An electronic circuit is a complete course of conductors through which current can travel. Circuits provide a path for current to flow. To be a circuit, this path must start and end at the same point. In other words, a circuit must form a loop. An electronic circuit is composed of individual electronic components, such as resistors, transistors, capacitors, inductors and diodes, connected by conductive wires or traces through which electric current can flow. The combination of components and wires allows various simple and complex operations to be performed: signals can be amplified, computations can be performed, and data can be moved from one place to another. Circuits can be constructed of discrete components connected by individual pieces of wire, but today it is much more common to create interconnections by photolithographic techniques on a laminated substrate and solder the components to these interconnections to create a finished circuit. In an integrated circuit or IC, the components and interconnections are formed on the same substrate, typically a semiconductor such as silicon or (less commonly) gallium arsenide. To design any electrical circuit, either analog or digital, electrical engineers need to be able to predict the voltages and currents at all places within the circuit. Linear circuits, that is, circuits wherein the outputs are linearly dependent on the inputs, can be analyzed by hand using complex analysis. Simple nonlinear circuits can also be analyzed in this way. Specialized software has been created to analyze circuits that are either too complicated or too nonlinear to analyze by hand.

Content Coverage

Chapters One and *Two* present the information about the diodes and transistors, correspondingly. Semiconductor diodes were the first semiconductor electronic devices. Transistors, however, are the basic building blocks of microchips, and roughly define the difference between electric and electronic devices.

Chapters Three and *Four* signify the theory of capacitors and MOSFET circuits, consistently. The physical form and construction of practical capacitors vary widely and many capacitor types are in common use. A MOSFET amplifier is a subcategory of amplifiers that employs metal–oxide–semiconductor field-effect transistor (MOSFET) technology to process digital signals with relatively low power consumption.

Chapter Five encompasses information about electronic oscillator that produces a periodic, oscillating electronic signal, often a sine wave or a square wave.

Chapter Six emphases on integrated circuit, or IC, that can function as an amplifier, oscillator, timer, microprocessor, or even computer memory. An IC is a small wafer, usually made of silicon, that can hold anywhere from hundreds to millions of transistors, resistors, and capacitors.

Chapters Seven and *Eight* are envisioned to focus on telecommunications systems and the need of instrumentation systems, separately. Telecommunications systems of today are transferred through many mediums such as the telephone, radio, television, LAN networks and the ever-growing internet. The basic need of instrumentation in a process is to get the best and most amount of information so as to successfully complete the process.

Chapter Nine clarifies about the MicroElectroMechanical systems (MEMS), which are made up of components between 1 and 100 micrometres in size. The fabrication of MEMS evolved from the process technology in semiconductor device fabrication, i.e. the basic techniques are deposition of material layers, patterning by photolithography and etching to produce the required shapes.

CHAPTER 1

Introduction to Diode

Chapter Contents

- **Introduction**
- **Main Functions**
- **History**
- **Thermionic Diodes**
- **Semiconductor Diodes**
- **Types of Semiconductor Diode**
- **Applications**

A diode is a specialized electronic component with two electrodes called the anode and the cathode. Most diodes are made with semiconductor materials such as silicon, germanium, or selenium. Some diodes are comprised of metal electrodes in a chamber evacuated or filled with a pure elemental gas at low pressure. Diodes can be used as rectifiers, signal limiters, voltage regulators, switches, signal modulators, signal mixers, signal demodulators, and oscillators.

The fundamental property of a diode is its tendency to conduct electric current in only one direction. When the cathode is negatively charged relative to the anode at a voltage greater than a certain minimum called forward breakover, then current flows through the diode. If the cathode is positive with respect to the anode, is at the same voltage as the anode, or is negative by an amount less than the forward breakover voltage, then the diode does not conduct current. This is a simplistic view, but is true for diodes operating as rectifiers, switches, and limiters. The forward breakover voltage is approximately six tenths of a volt (0.6 V) for silicon devices, 0.3 V for germanium devices, and 1 V for selenium devices.

INTRODUCTION

In electronics, a diode is a two-terminal electronic component that conducts primarily in one direction (asymmetric conductance); it has low (ideally zero) resistance to the flow of current in one direction, and high (ideally infinite) resistance in the other. A semiconductor diode, the most common type today, is a crystalline piece of semiconductor material with a p–n junction connected to two electrical terminals. A vacuum tube diode has two electrodes, a plate (anode) and a heated cathode. Semiconductor diodes were the first semiconductor electronic devices. The discovery of crystals' rectifying abilities was made by German physicist Ferdinand Braun in 1874. The first semiconductor diodes, called cat's whisker diodes, developed around 1906, were made of mineral crystals such as galena. Today, most diodes are made of silicon, but other semiconductors such as selenium or germanium are sometimes used.

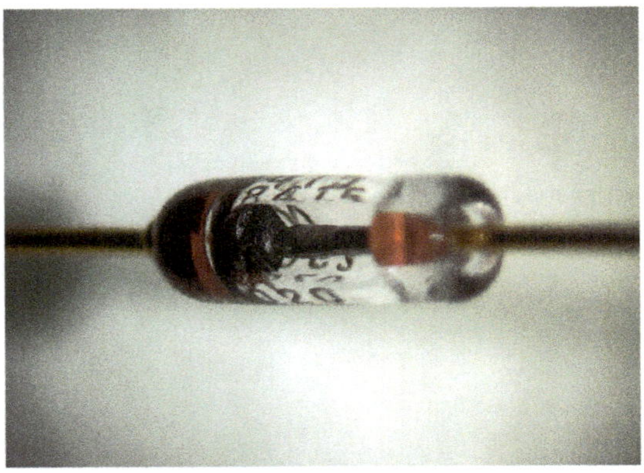

Closeup of a diode, showing the square-shaped semiconductor crystal (black object on left).

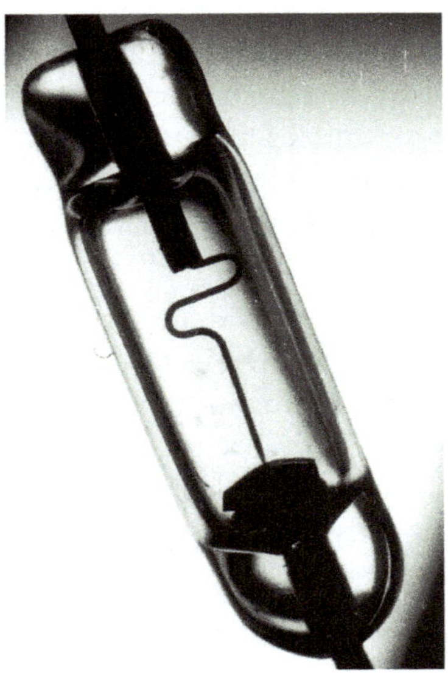

Extreme macro photo of a Chinese diode of the seventies.

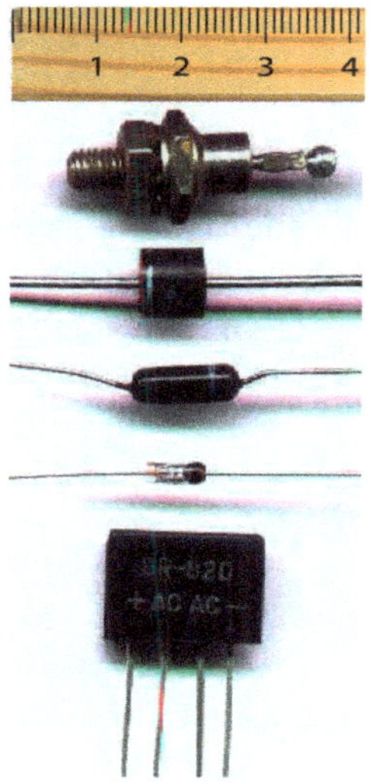

Various semiconductor diodes. Bottom: A bridge rectifier. In most diodes, a white or black painted band identifies the cathode into which electrons will flow when the diode is conducting. Electron flow is the reverse of conventional current flow.

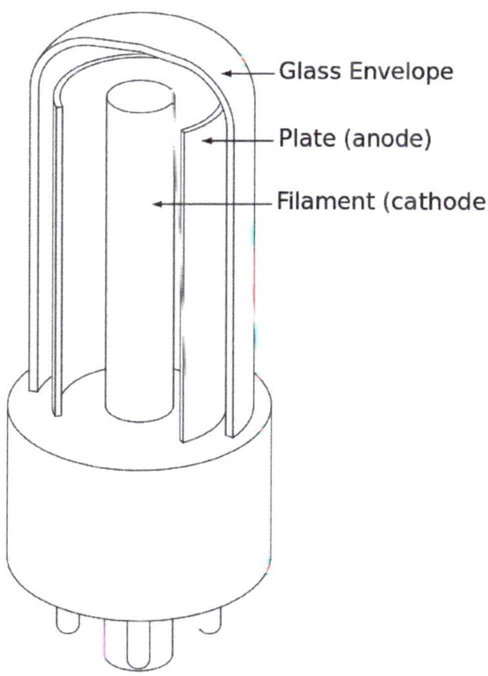

Structure of a vacuum tube diode. The filament may be bare, or more commonly (as shown here), embedded within and insulated from an enclosing cathode.

MAIN FUNCTIONS

The most common function of a diode is to allow an electric current to pass in one direction (called the diode's forward direction), while blocking current in the opposite direction (the reverse direction). Thus, the diode can be viewed as an electronic version of a check valve. This unidirectional behavior is called rectification, and is used to convert alternating current (AC) to direct current (DC), including extraction of modulation from radio signals in radio receivers—these diodes are forms of rectifiers.

However, diodes can have more complicated behavior than this simple on–off action, because of their nonlinear current-voltage characteristics. Semiconductor diodes begin conducting electricity only if a certain threshold voltage or cut-in voltage is present in the forward direction (a state in which the diode is said to be forward-biased). The voltage drop across a forward-biased diode varies only a little with the current, and is a function of temperature; this effect can be used as a temperature sensor or as a voltage reference.

A semiconductor diode's current–voltage characteristic can be tailored by selecting the semiconductor materials and the doping impurities introduced into the materials during manufacture. These techniques are used to create special-purpose diodes that perform many different functions. For example, diodes are used to regulate voltage (Zener diodes), to protect circuits from high voltage surges (avalanche diodes), to electronically tune radio and TV receivers (varactor diodes), to generate radio-frequency oscillations (tunnel diodes, Gunn diodes, IMPATT diodes), and to produce light (light-emitting diodes). Tunnel, Gunn and IMPATT diodes exhibit negative resistance, which is useful in microwave and switching circuits.

Diodes, both vacuum and semiconductor, can be used as shot-noise generators.

HISTORY

Thermionic (vacuum tube) diodes and solid state (semiconductor) diodes were developed separately, at approximately the same time, in the early 1900s, as radio receiver detectors. Until the 1950s vacuum tube diodes were used more frequently in radios because the early point-contact type semiconductor diodes were less stable. In addition, most receiving sets had vacuum tubes for amplification that could easily have the thermionic diodes included in the tube (for example the 12SQ7 double diode triode), and vacuum tube rectifiers and gas-filled rectifiers were capable of handling some high voltage/high current rectification tasks better than the semiconductor diodes (such as selenium rectifiers) which were available at that time.

Vacuum Tube Diodes

In 1873, Frederick Guthrie discovered the basic principle of operation of thermionic diodes. Guthrie discovered that a positively charged electroscope could be discharged by bringing a grounded piece of white-hot metal close to it (but not actually touching it). The same did not apply to a negatively charged electroscope, indicating that the current flow was only possible in one direction.

Thomas Edison independently rediscovered the principle on February 13, 1880. At the time, Edison was investigating why the filaments of his carbon-filament light bulbs nearly always burned out at the positive-connected end. He had a special bulb made with a metal plate sealed into the glass envelope. Using this device, he confirmed that an invisible current flowed from the glowing filament through the vacuum to the metal plate, but only when the plate was connected to the positive supply.

Edison devised a circuit where his modified light bulb effectively replaced the resistor in a DC voltmeter. Edison was awarded a patent for this invention in 1884. Since there was no apparent practical use for such a device at the time, the patent application was most likely simply a precaution in case someone else did find a use for the so-called Edison effect.

About 20 years later, John Ambrose Fleming (scientific adviser to the Marconi Company and former Edison employee) realized that the Edison effect could be used as a precision radio detector. Fleming patented the first true thermionic diode, the Fleming valve, in Britain on November 16, 1904 (followed by U.S. Patent 803,684 in November 1905).

Solid-State Diodes

In 1874 German scientist Karl Ferdinand Braun discovered the "unilateral conduction" of crystals. Braun patented the crystal rectifier in 1899. Copper oxide and selenium rectifiers were developed for power applications in the 1930s.

Indian scientist Jagadish Chandra Bose was the first to use a crystal for detecting radio waves in 1894. The crystal detector was developed into a practical device for wireless telegraphy by Greenleaf Whittier Pickard, who invented a silicon crystal detector in 1903 and received a patent for it on November 20, 1906. Other experimenters tried a variety of other substances, of which the most widely used was the mineral galena (lead sulfide). Other substances offered slightly better performance, but galena was most widely used because it had the advantage of being cheap and easy to obtain. The crystal detector in these early crystal radio sets consisted of an adjustable wire point-contact, often made of gold or platinum because of their incorrodible nature (the so-called "cat's whisker"), which could be manually moved over the face of the crystal in search of a portion of that mineral with rectifying qualties. This troublesome device was superseded by thermionic diodes (vacuum tubes) by the 1920s, but after high purity semiconductor materials became available, the crystal detector returned to dominant use with the advent, in the 1950s, of inexpensive fixed-germanium diodes. Bell Labs also developed a germanium diode for microwave reception, and AT&T used these in their microwave towers that criss-crossed the nation starting in the late 1940s, carrying telephone and network television signals. Bell Labs did not develop a satisfactory thermionic diode for microwave reception.

Etymology

At the time of their invention, such devices were known as rectifiers. In 1919, the year tetrodes were invented, William Henry Eccles coined the term ***diode*** from the Greek roots di (from *δί*), meaning 'two', and ode, meaning 'path'. (However, the word diode itself, as well as triode, tetrode, pentode, hexode, were already in use as terms of multiplex telegraphy.

Rectifiers

Although all diodes rectify, the term 'rectifier' is normally reserved for higher currents and voltages than would normally be found in the rectification of lower power signals; examples include:

- Power supply rectifiers (half-wave, full-wave, bridge)
- Flyback diodes

World's Smallest Diode

Researchers from the University of Georgia and Ben-Gurion University of the Negev (BGU) have developed a diode made from a molecule of DNA. Professor Bingqian Xu from the College of Engineering at the University of Georgia and his team took a single DNA molecule made from 11 base pairs and connected it to an electronic circuit a few nanometers in size. When layers of coralyne were inserted between layers of DNA, the current jumped up to 15 times larger negative versus positive, which is necessary for a nano diode.

THERMIONIC DIODES

A thermionic diode is a thermionic-valve device (also known as a vacuum tube, tube, or valve), consisting of a sealed evacuated glass envelope containing two electrodes: a cathode heated by a filament, and a plate (anode). Early examples were fairly similar in appearance to incandescent light bulbs.

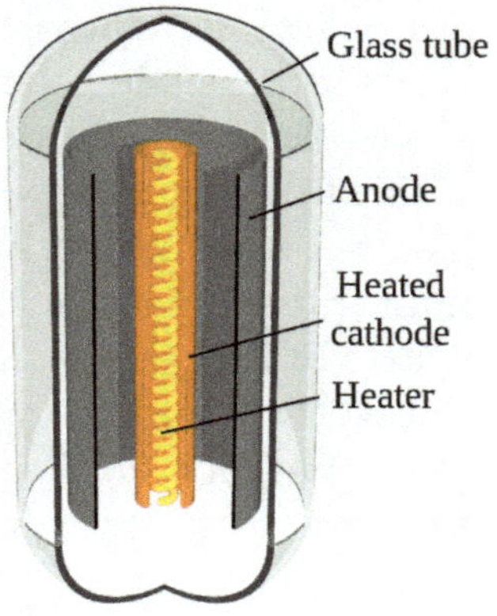

Diode vacuum tube construction

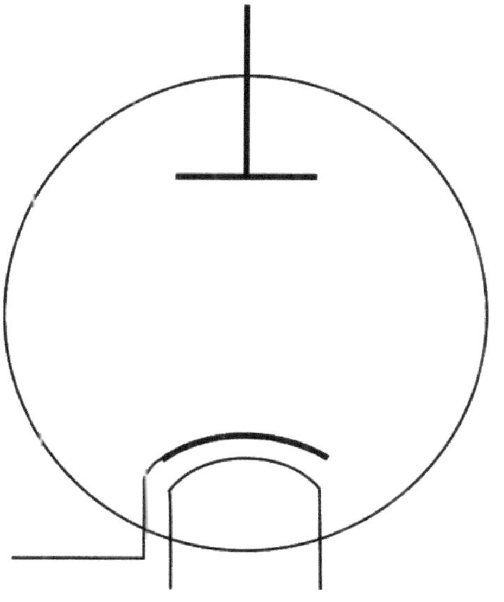

The symbol for an indirect heated vacuum-tube diode. From top to bottom, the components are the anode, the cathode, and the heater filament.

In operation, a current flows through the filament (heater)—a high resistance wire made of nichrome—and heats the cathode red hot (800–1000 °C). This causes the cathode to releases electrons into the vacuum, a process called thermionic emission. (Some valves use direct heating, in which a tungsten filament acts as both heater and cathode.) The alternating voltage to be rectified is applied between the cathode and the concentric plate electrode. When the plate has a positive voltage with respect to the cathode, it electrostatically attracts the electrons from the cathode, so a current of electrons flows through the tube from cathode to plate. However, when the polarity is reversed and the plate has a negative voltage, no current flows, because the cathode electrons are not attracted to it. The plate, being unheated, does not emit any electrons. So electrons can only flow through the tube in one direction, from the cathode to the anode plate.

The cathode is coated with oxides of alkaline earth metals, such as barium and strontium oxides. These have a low work function, meaning that they more readily emit electrons than would the uncoated cathode.

In a mercury-arc valve, an arc forms between a refractory conductive anode and a pool of liquid mercury acting as cathode. Such units were made with ratings up to hundreds of kilowatts, and were important in the development of HVDC power transmission. Some types of smaller thermionic rectifiers had mercury vapor fill to reduce their forward voltage drop and to increase current rating over thermionic hard-vacuum devices.

Throughout the vacuum tube era, valve diodes were used in analog signal applications and as rectifiers in DC power supplies in consumer electronics such as radios, televisions, and sound systems. They were replaced in power supplies beginning in the 1940s by selenium rectifiers and then by semiconductor diodes by the 1960s. Today they are still used in a few high power applications where their ability to withstand transient voltages and their robustness gives them an advantage over semiconductor devices. The recent (2012) resurgence of interest among audiophiles and recording studios in old valve audio gear such as guitar amplifiers and home audio systems has provided a market for the legacy consumer diode valves.

SEMICONDUCTOR DIODES

Electronic Symbols

The symbol used for a semiconductor diode in a circuit diagram specifies the type of diode. There are alternative symbols for some types of diodes, though the differences are minor. The triangle in the symbols points to the forward direction.

Diode

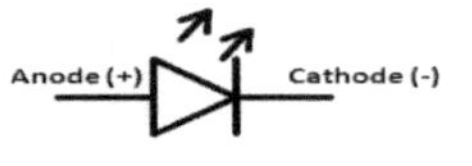

Light-emitting diode (LED)

Photodiode

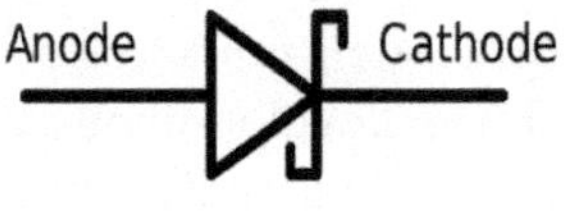

Schottky diode

Transient-voltage-suppression diode (TVS)

Tunnel diode

Varicap

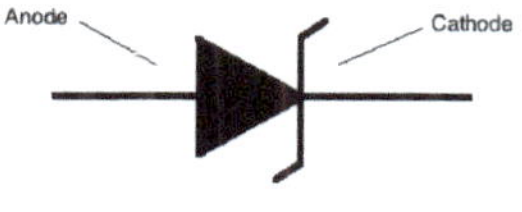

Zener diode

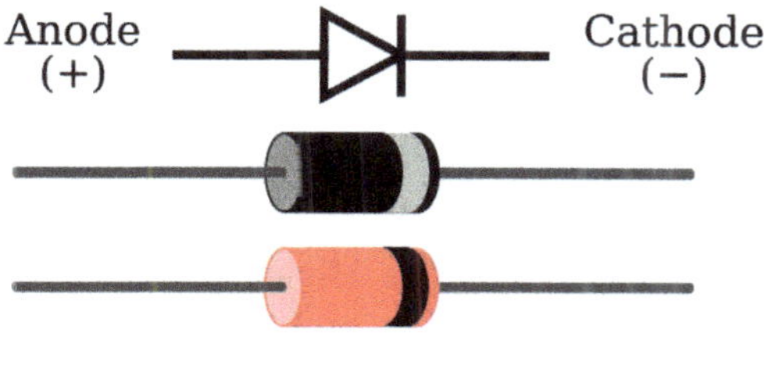

Typical diode packages in same alignment as diode symbol. Thin bar depicts the cathode.

A galena cat's-whisker detector, a point-contact diode.

Point-Contact Diodes

A point-contact diode works the same as the junction diodes described below, but its construction is simpler. A pointed metal wire is placed in contact with an n-type semiconductor. Some metal migrates into the semiconductor to make a small p-type region around the contact. The 1N34 germanium version is still used in radio receivers as a detector and occasionally in specialized analog electronics.

Junction Diodes

p–n Junction Diode

A p–n diode is a type of semiconductor diode based upon the p–n junction. The diode conducts current in only one direction, and it is made by joining a p-type semiconducting layer to an n-type semiconducting layer. Semiconductor diodes have multiple uses including rectification of alternating current to direct current, detection of radio signals, emitting light and detecting light.

Structure

The figure shows two of the many possible structures used for p–n semiconductor diodes, both adapted to increase the voltage the devices can withstand in reverse bias. The top structure uses a mesa to avoid a sharp curvature of the p^+-region next to the adjoining n-layer. The bottom structure uses a lightly doped p-guard-ring at the edge of the sharp corner of the p^+-layer to spread the voltage out over a larger distance and reduce the electric field. (Superscripts like n^+ or n^- refer to heavier or lighter impurity doping levels.)

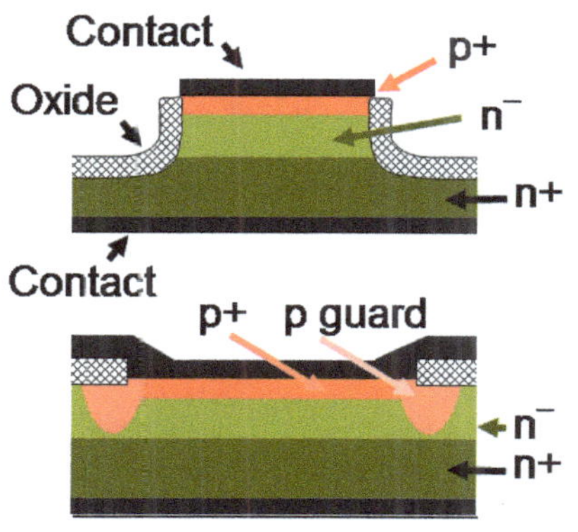

Mesa diode structure (top) and planar diode structure with guard-ring (bottom).

Electrical Behavior

The ideal diode has zero resistance for the forward bias polarity, and infinite resistance (conducts zero current) for the reverse voltage polarity; if connected in an alternating current circuit, the semiconductor diode acts as an electrical rectifier.

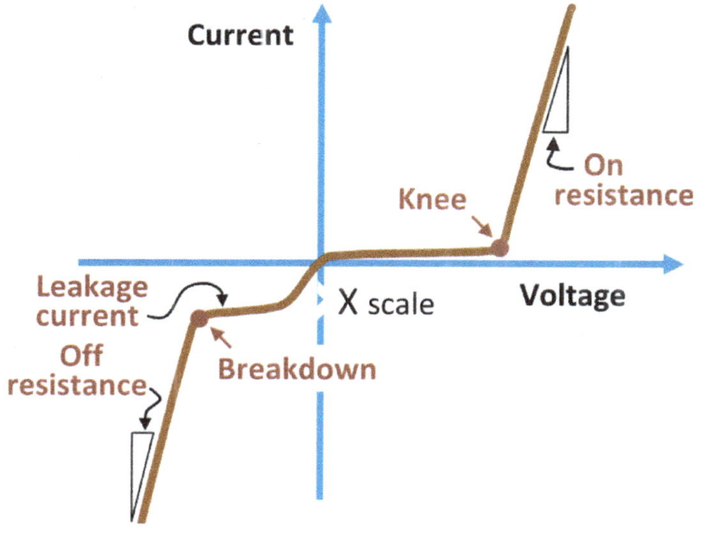

Nonideal p–n diode current-voltage characteristics.

The semiconductor diode is not ideal. As shown in the figure, the diode does not conduct appreciably until a nonzero knee voltage (also called the turn-on voltage or the cut-in voltage) is reached. Above this voltage the slope of the current-voltage curve is not infinite (on-resistance is not zero). In the reverse direction the diode conducts a nonzero leakage current (exaggerated by a smaller scale in the figure) and at a sufficiently large reverse voltage below the breakdown voltage the current increases very rapidly with more negative reverse voltages.

As shown in the figure, the on and off resistances are the reciprocal slopes of the current-voltage characteristic at a selected bias point:

$$r_D = \left.\frac{\Delta v_D}{\Delta i_D}\right|_{v_D=V_{BIAS}},$$

where r_D is the resistance and Δi_D is the current change corresponding to the diode voltage change Δv_D at the bias $v_D=V_{BIAS}$.

Operation

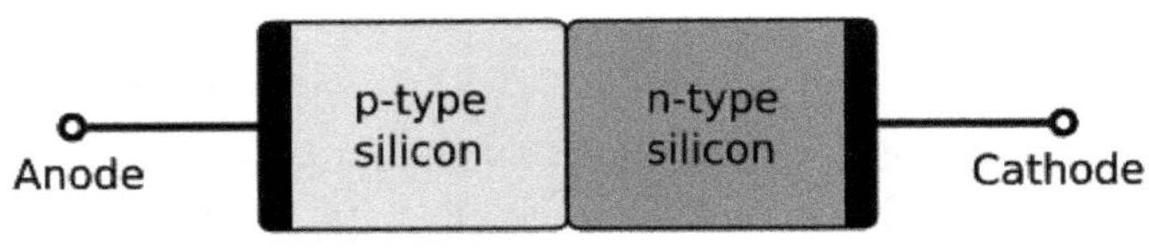

An abrupt p–n diode made by doping silicon.

Here, the operation of the abrupt p–n diode is considered. By "abrupt" is meant that the p- and n-type doping exhibit a step function discontinuity at the plane where they encounter each other. The objective is to explain the various bias regimes in the figure displaying current-voltage characteristics. Operation is described using band-bending diagrams that show how the lowest conduction band energy and the highest valence band energy vary with position inside the diode under various bias conditions.

Zero Bias

The figure shows a band bending diagram for a p–n diode; that is, the band edges for the conduction band (upper line) and the valence band (lower line) are shown as a function of position on both sides of the junction between the p-type material (left side) and the n-type material (right side). When a p-type and an n-type region of the same semiconductor are brought together and the two diode contacts are short-circuited, the Fermi half-occupancy level (dashed horizontal straight line) is situated at a constant level. This level ensures that in the field-free bulk on both sides of the junction the hole and electron occupancies are correct. (So, for example, it is not necessary for an electron to leave the n-side and travel to the p-side through the short circuit to adjust the occupancies.)

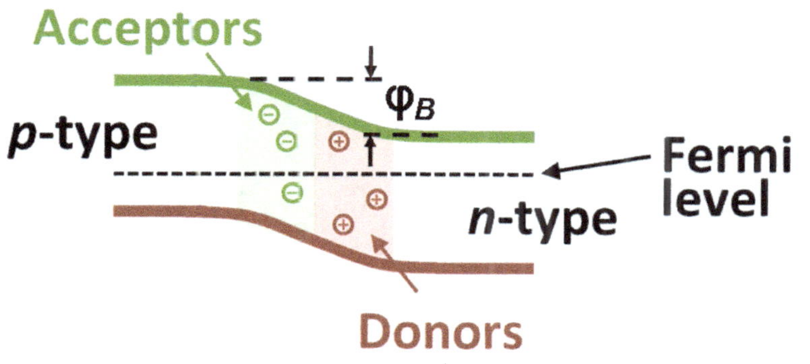

Band-bending diagram for p–n diode at zero applied voltage. The depletion region is shaded.

However, a flat Fermi level requires the bands on the p-type side to move higher than the corresponding bands on the n-type side, forming a step or barrier in the band edges, labeled φ_B. This step forces the electron density on the p-side to be a Boltzmann factor $\exp(-\varphi_B/V_{th})$ smaller than on the n-side, corresponding to the lower electron density in p-region. The symbol V_{th} denotes the thermal voltage, defined as $V_{th} = k_BT/q$. At T = 290 kelvins (room temperature), the thermal voltage is approximately 25 mV. Similarly, hole density on the n-side is a Boltzmann factor smaller than on the p-side. This reciprocal reduction in minority carrier density across the junction forces the pn-product of carrier densities to be

$$pn = \left(p_B n_B \, e^{-\varphi_B/V_{th}}\right) \, e^{v_D/V_{th}} \, .$$

at any position within the diode at equilibrium. Where p_B and n_B are the bulk majority carrier densities on the p-side and the n-side, respectively.

As a result of this step in band edges, a depletion region near the junction becomes depleted of both holes and electrons, forming an insulating region with almost no mobile charges. There are, however, fixed, immobile charges due to dopant ions. The near absence of mobile charge in the depletion layer means that the mobile charges present are insufficient to balance the immobile charge contributed by the dopant ions: a negative charge on the p-type side due to acceptor dopant and as a positive charge on the n-type side due to donor dopant. Because of this charge there is an electric field in this region, as determined by Poisson's equation. The width of the depletion region adjusts so the negative acceptor charge on the p-side exactly balances the positive donor charge on the n-side, so there is no electric field outside the depletion region on either side.

In this band configuration no voltage is applied and no current flows through the diode. To force current through the diode a forward bias must be applied, as described next.

Forward Bias

In forward bias, positive terminal of the battery is connected to the p- type material and negative terminal is connected to the n- type material so that holes are injected into the p-type material and electrons into the n-type material. The electrons in the n-type material are called majority carriers on that side, but electrons that make it to the p-type side are called minority carriers. The same descriptors apply to holes: they are majority carriers on the p-type side, and minority carriers on the n-type side.

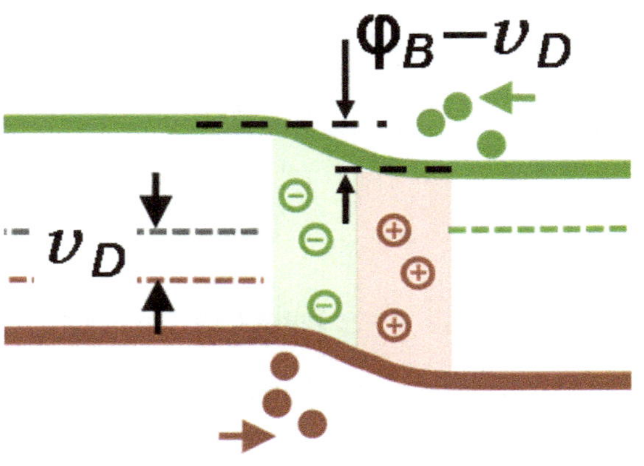

Band-bending diagram for p–n diode in forward bias. Diffusion drives carriers across the junction.

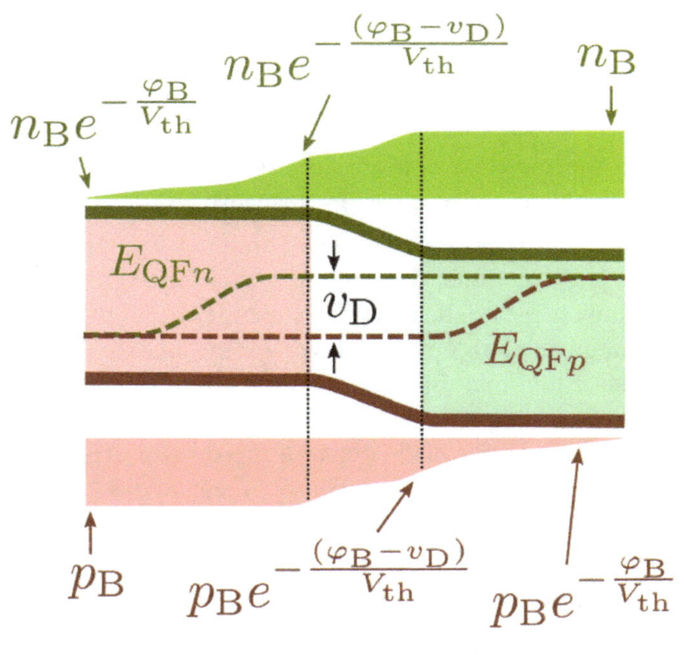

Quasi-Fermi levels and carrier densities in forward biased p–n- diode. The figure assumes recombination is confined to the regions where majority carrier concentration is near the bulk values, which is not accurate when recombination-generation centers in the field region play a role.

A forward bias separates the two bulk half-occupancy levels by the amount of the applied voltage, which lowers the separation of the p-type bulk band edges to be closer in energy to those of the n-type. As shown in the diagram, the step in band edges is reduced by the applied voltage to $\varphi_B - v_D$. (The band bending diagram is made in units of volts, so no electron charge appears to convert v_D to energy.)

Under forward bias, a diffusion current flows (that is a current driven by a concentration gradient) of holes from the p-side into the n-side, and of electrons in the opposite direction from the n-side to the p-side. The gradient driving this transfer is set up as follows: in the bulk distant from the interface, minority carriers have a very low concentration compared to majority carriers, for example, electron density on the p-side (where they are minority carriers) is a factor $\exp(-\varphi_B/V_{th})$ lower than on the n-side (where they are majority carriers). On the other hand, near the interface, application of voltage v_D reduces the step in band edges and increases minority carrier densities by a Boltzmann factor $\exp(v_D/V_{th})$ above the bulk values. Within the junction, the pn-product is increased above the equilibrium value to:

$$pn = \left(p_B n_B\, e^{-\varphi_B/V_{th}}\right)\, e^{v_D/V_{th}}\ .$$

The gradient driving the diffusion is then the difference between the large excess minority carrier densities at the barrier and the low densities in the bulk, and that gradient drives diffusion of minority carriers from the interface into the bulk. The injected minority carriers are reduced in number as they travel into the bulk by recombination mechanisms that drive the excess concentrations toward the bulk values.

Recombination can occur by direct encounter with a majority carrier, annihilating both carriers, or through a recombination-generation center, a defect that alternately traps holes and electrons, assisting recombination. The minority carriers have a limited lifetime, and this lifetime in turn limits how far they can diffuse from the majority carrier side into the minority carrier side, the so-called diffusion length. In the LED recombination of electrons and holes is accompanied by emission of light of a wavelength related to the energy gap between valence and conduction bands, so the diode converts a portion of the forward current into light.

Under forward bias, the half-occupancy lines for holes and electrons cannot remain flat throughout the device as they are when in equilibrium, but become quasi-Fermi levels that vary with position. As shown in the figure, the electron quasi-Fermi level shifts with position, from the half-occupancy equilibrium Fermi level in the n-bulk, to the half-occupancy equilibrium level for holes deep in the p-bulk. The hole quasi-Fermi level does the reverse. The two quasi-Fermi levels do not coincide except deep in the bulk materials.

The figure shows the majority carrier densities drop from the majority carrier density levels n_B, p_B in their respective bulk materials, to a level a factor $\exp(-(\varphi_B - v_D)/V_{th})$ smaller at the top of the barrier, which is reduced from the equilibrium value φ_B by the amount of the forward diode bias v_D. Because this barrier is located in the oppositely doped material, the injected carriers at the barrier position are now minority carriers. As recombination takes hold, the minority carrier densities drop with depth to their equilibrium values for bulk minority carriers, a factor $\exp(-\varphi_B/V_{th})$ smaller than their bulk densities n_B, p_B as majority carriers before injection. At this point the quasi-Fermi levels rejoin the bulk Fermi level positions.

The reduced step in band edges also means that under forward bias the depletion region narrows as holes are pushed into it from the p-side and electrons from the n-side.

In the simple p–n diode the forward current increases exponentially with forward bias voltage due to the exponential increase in carrier densities, so there is always some current at even very small values of applied voltage. However, if one is interested in some particular current level, it will require a "knee" voltage before that current level is reached. For example, a very common choice in texts about circuits using silicon diodes is $V_{Knee} = 0.7$ V. Above the knee, the current continues to increase exponentially. Some special diodes, such as some varactors, are designed deliberately to maintain a low current level up to some knee voltage in the forward direction.

Reverse Bias

In reverse bias the occupancy level for holes again tends to stay at the level of the bulk p-type semiconductor while the occupancy level for electrons follows that for the bulk n-type. In this case, the p-type bulk band edges are raised relative to the n-type bulk by the reverse bias v_R, so the two bulk occupancy levels are separated again by an energy determined by the applied voltage. As shown in the diagram, this behavior means the step in band edges is increased to φ_B+v_R, and the depletion region widens as holes are pulled away from it on the p-side and electrons on the n-side.

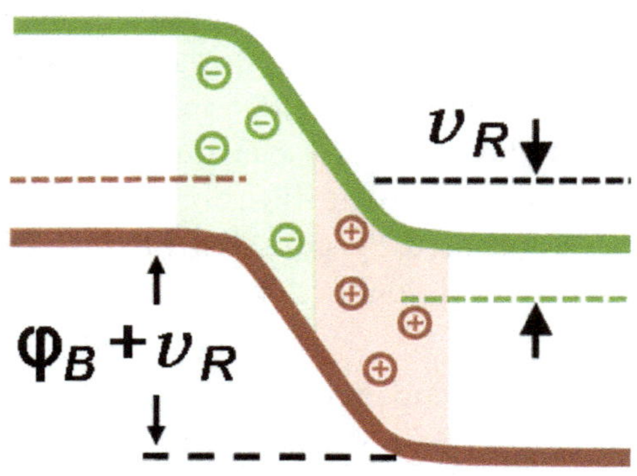

Band-bending for p–n diode in reverse bias

When the reverse bias is applied, the electric field in the depletion region is increased, pulling the electrons and holes further apart than in the zero bias case. Thus, any current that flows is due to the very weak process of carrier generation inside the depletion region due to generation-recombination defects in this region. That very small current is the source of the leakage current under reverse bias. In the photodiode, reverse current is introduced using creation of holes and electrons in the depletion region by incident light, thus converting a portion of the incident light into an electric current.

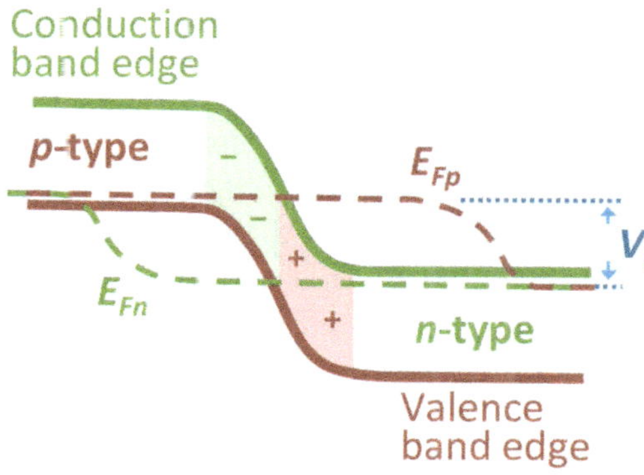

Quasi-Fermi levels in reverse-biased p–n diode.

When the reverse bias becomes very large, reaching the breakdown voltage, the generation process in the depletion region accelerates leading to an avalanche condition which can cause runaway and destroy the diode.

Diode Law

The DC current-voltage behavior of the ideal p–n diode is governed by the Shockley diode equation:

$$i_{\rm D} = I_{\rm R}\left(e^{v_{\rm D}/V_{\rm th}} - 1\right),$$

where v_D is the DC voltage across the diode and I_R is the reverse saturation current, the current that flows when the diode is reverse biased (that is, v_D is large and negative). The quantity V_{th} is the thermal voltage defined as $V_{th} = k_BT/q$. This is approximately equal to 25 mV at T = 290 kelvins.

This equation does not model the non-ideal behavior such as excess reverse leakage or breakdown phenomena. In many practical diodes this equation must be modified to read

$$i_D = I_R\left(e^{v_{\rm D}/nV_{\rm th}} - 1\right),$$

where n is an ideality factor introduced to model a slower rate of increase than predicted by the ideal diode law. Using this equation, the diode on-resistance is

$$r_{\rm D} = \frac{1}{di_{\rm D}/dv_{\rm D}} \approx \frac{nV_{\rm th}}{i_{\rm D}},$$

exhibiting a lower resistance the higher the current.

Capacitance

The depletion layer between the n- and p-sides of a p–n-diode serves as an insulating region that separates the two diode contacts. Thus, the diode in reverse bias exhibits a depletion-layer capacitance, sometimes more vaguely called a junction capacitance, analogous to a parallel plate capacitor with a dielectric spacer between the contacts. In reverse bias the width of the depletion layer is widened with increasing reverse bias v_R, and the capacitance is accordingly decreased. Thus, the junction serves as a voltage-controllable capacitor. In a simplified one-dimensional model, the junction capacitance is:

$$C_J = \kappa\varepsilon_0 \frac{A}{w(v_R)} \ ,$$

with A the device area, κ the relative semiconductor dielectric permittivity, ε_0 the electric constant, and w the depletion width (thickness of the region where mobile carrier density is negligible).

In forward bias, besides the above depletion-layer capacitance, minority carrier charge injection and diffusion occurs. A diffusion capacitance exists expressing the change in minority carrier charge that occurs with a change in forward bias. In terms of the stored minority carrier charge, the diode current i_D is:

$$i_D = \frac{Q_D}{\tau_T} \ ,$$

where Q_D is the charge associated with diffusion of minority carriers, and τ_T is the transit time, the time taken for the minority charge to transit the injection region. Typical values for transit time are 0.1–100 ns. On this basis, the diffusion capacitance is calculated to be:

$$C_D = \frac{dQ_D}{dv_D} = \tau_T \frac{di_D}{dv_D} = \frac{i_D \tau_T}{V_{th}} \ .$$

Generally speaking, for usual current levels in forward bias, this capacitance far exceeds the depletion-layer capacitance.

Transient Response

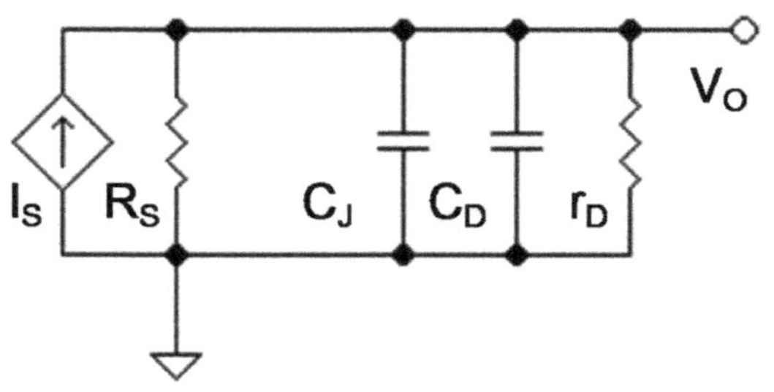

Small-signal circuit for p–n diode driven by a current signal represented as a Norton source.

The diode is a highly non-linear device, but for small-signal variations its response can be analyzed using a small-signal circuit based upon the DC bias about which the signal is imagined to vary. The equivalent circuit is shown at the right for a diode driven by a Norton source. Using Kirchhoff's current law at the output node:

$$I_S = \left(j\omega(C_J + C_D) + \frac{1}{r_D} + \frac{1}{R_S} \right) V_O \,,$$

with C_D the diode diffusion capacitance, C_J the diode junction capacitance (the depletion layer capacitance) and r_D the diode resistance, all at the selected quiescent bias point or Q-point. The output voltage provided by this circuit is then:

$$\frac{V_O}{I_S} = \frac{(R_S \| r_D)}{1 + j\omega(C_D + C_J)(R_S \| r_D)} \,,$$

with $(R_S||r_D)$ the parallel combination of R_S and r_D. This transresistance amplifier exhibits a corner frequency, denoted f_C:

$$f_C = \frac{1}{2\pi(C_D + C_J)(R_S \| r_D)} \,,$$

and for frequencies $f >> f_C$ the gain rolls off with frequency as the capacitors short-circuit the resistor r_D. Assuming, as is the case when the diode is turned on, that $C_D >> C_J$ and $R_S >> r_D$, the expressions found above for the diode resistance and capacitance provide:

$$f_C = \frac{1}{2\pi n \tau_T} \,,$$

which relates the corner frequency to the diode transit time τ_T.

For diodes operated in reverse bias, C_D is zero and the term corner frequency often is replaced by cutoff frequency. In any event, in reverse bias the diode resistance becomes quite large, although not infinite as the ideal diode law suggests, and the assumption that it is less than the Norton resistance of the driver may not be accurate. The junction capacitance is small and depends upon the reverse bias v_R. The cutoff frequency is then:

$$f_C = \frac{1}{2\pi C_J (R_S \| r_D)} \,,$$

and varies with reverse bias because the width $w(v_R)$ of the insulating region depleted of mobile carriers increases with increasing diode reverse bias, reducing the capacitance.

Schottky Diode

The Schottky diode (named after German physicist Walter H. Schottky), also known as hot carrier diode, is a semiconductor diode formed by the junction of a semiconductor with a metal. It has a low forward voltage drop and a very fast switching action. The cat's-whisker detectors used in the early days of wireless and metal rectifiers used in early power applications can be considered primitive Schottky diodes.

When sufficient forward voltage is applied, a current flows in the forward direction. A silicon diode has a typical forward voltage of 600–700 mV, while the Schottky's forward voltage is 150 – 450 mV. This lower forward voltage requirement allows higher switching speeds and better system efficiency.

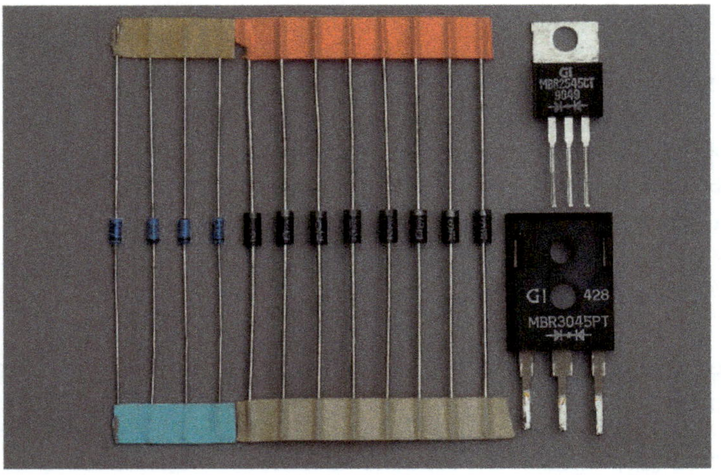

Various Schottky barrier diodes: Small signal RF devices (left), medium and high power Schottky rectifying diodes (middle and right).

Construction

A metal–semiconductor junction is formed between a metal and a semiconductor, creating a Schottky barrier (instead of a semiconductor–semiconductor junction as in conventional diodes). Typical metals used are molybdenum, platinum, chromium or tungsten, and certain silicides (e.g., palladium silicide and platinum silicide), whereas the semiconductor would typically be n-type silicon. The metal side acts as the anode, and n-type semiconductor acts as the cathode of the diode. This Schottky barrier results in both very fast switching and low forward voltage drop.

1N5822 Schottky diode with cut-open packaging. The semiconductor in the center makes a Schottky barrier against one metal electrode (providing rectifying action) and an ohmic contact with the other electrode.

The choice of the combination of the metal and semiconductor determines the forward voltage of the diode. Both n- and p-type semiconductors can develop Schottky barriers. However, the p-type typically has a much lower forward voltage. As the reverse leakage current increases dramatically with lowering the forward voltage, it can not be too low, so the usually employed range is about 0.5–0.7 V, and p-type semiconductors are employed only rarely. Titanium silicide and other refractory silicides, which are able to withstand the temperatures needed for source/drain annealing in CMOS processes, usually have too low a forward voltage to be useful, so processes using these silicides therefore usually do not offer Schottky diodes.

With increased doping of the semiconductor, the width of the depletion region drops. Below a certain width, the charge carriers can tunnel through the depletion region. At very high doping levels, the junction does not behave as a rectifier anymore and becomes an ohmic contact. This can be used for the simultaneous formation of ohmic contacts and diodes, as a diode will form between the silicide and lightly doped n-type region, and an ohmic contact will form between the silicide and the heavily doped n- or p-type region. Lightly doped p-type regions pose a problem, as the resulting contact has too high a resistance for a good ohmic contact, but too low a forward voltage and too high a reverse leakage to make a good diode.

As the edges of the Schottky contact are fairly sharp, a high electric field gradient occurs around them, which limits how large the reverse breakdown voltage threshold can be. Various strategies are used, from guard rings to overlaps of metallization to spread out the field gradient. The guard rings consume valuable die area and are used primarily for larger higher-voltage diodes, while overlapping metallization is employed primarily with smaller low-voltage diodes.

Schottky diodes are often used as antisaturation clamps in Schottky transistors. Schottky diodes made from palladium silicide (PdSi) are excellent due to their lower forward voltage (which has to be lower than the forward voltage of the base-collector junction). The Schottky temperature coefficient is lower than the coefficient of the B–C junction, which limits the use of PdSi at higher temperatures. For power Schottky diodes, the parasitic resistances of the buried n+ layer and the epitaxial n-type layer become important. The resistance of the epitaxial layer is more important than it is for a transistor, as the current must cross its entire thickness. However, it serves as a distributed ballasting resistor over the entire area of the junction and, under usual conditions, prevents localized thermal runaway.

In comparison with the power p–n diodes the Schottky diodes are less rugged. The junction is direct contact with the thermally sensitive metallization, a Schottky diode can therefore dissipate less power than an equivalent-size p-n counterpart with a deep-buried junction before failing (especially during reverse breakdown). The relative advantage of the lower forward voltage of Schottky diodes is diminished at higher forward currents, where the voltage drop is dominated by the series resistance.

Reverse Recovery Time

The most important difference between the p-n diode and the Schottky diode is the reverse recovery time (t_{rr}), when the diode switches from the conducting to the non-conducting state. In a p–n diode, the reverse recovery time can be in the order of several microseconds to less than 100 ns for fast diodes. Schottky diodes do not have a recovery time, as there is nothing to recover from (i.e., there is no charge carrier depletion region at the junction). The switching time is ~100 ps for the small-signal diodes, and up to tens of nanoseconds for special high-capacity power diodes. With p–n-junction switching, there is also a reverse recovery current, which in high-power semiconductors brings increased EMI noise. With Schottky diodes, switching is essentially "instantaneous" with only a slight capacitive loading, which is much less of a concern.

This "instantaneous" switching is not always the case. In higher voltage Schottky devices, in particular, the guard ring structure needed to control breakdown field geometry creates a parasitic p-n diode with the usual recovery time attributes. As long as this guard ring diode is not forward biased, it adds only capacitance. If the Schottky junction is driven hard enough however, the forward voltage eventually will bias both diodes forward and actual t_{rr} will be greatly impacted.

It is often said that the Schottky diode is a "majority carrier" semiconductor device. This means that if the semiconductor body is a doped n-type, only the n-type carriers (mobile electrons) play a significant role in normal operation of the device. The majority carriers are quickly injected into the conduction band of the metal contact on the other side of the diode to become free moving electrons. Therefore, no slow random recombination of n and p type carriers is involved, so that this diode can cease conduction faster than an ordinary p–n rectifier diode. This property in turn allows a smaller device area, which also makes for a faster transition. This is another reason why Schottky diodes are useful in switch-mode power converters: the high speed of the diode means that the circuit can operate at frequencies in the range 200 kHz to 2 MHz, allowing the use of small inductors and capacitors with greater efficiency than would be possible with other diode types. Small-area Schottky diodes are the heart of RF detectors and mixers, which often operate at frequencies up to 50 GHz.

Limitations

The most evident limitations of Schottky diodes are their relatively low reverse voltage ratings, and their relatively high reverse leakage current. For silicon-metal Schottky diodes, the reverse voltage is typically 50 V or less. Some higher-voltage designs are available (200 V is considered a high reverse voltage). Reverse leakage current, since it increases with temperature, leads to a thermal instability issue. This often limits the useful reverse voltage to well below the actual rating.

While higher reverse voltages are achievable, they would present a higher forward voltage, comparable to other types of standard diodes. Such Schottky diodes would have no advantage unless great switching speed is required.

Silicon Carbide Schottky Diode

Schottky diodes constructed from silicon carbide have a much lower reverse leakage current than silicon Schottky diodes, as well as higher forward voltage and reverse voltage. As of 2011 they were available from manufacturers in variants up to 1700 V of reverse voltage.

Silicon carbide has a high thermal conductivity, and temperature has little influence on its switching and thermal characteristics. With special packaging, silicon carbide Schottky diodes can operate at junction temperatures of over 500 K (about 200 °C), which allows passive radiative cooling in aerospace applications.

Applications

Voltage Clamping

While standard silicon diodes have a forward voltage drop of about 0.6 V and germanium diodes 0.2 V, Schottky diodes' voltage drop at forward biases of around 1 mA is in the range of 0.15 V to 0.46 V (see the 1N5817 and 1N5711 datasheets found online at manufacturer's websites), which makes them useful in voltage clamping applications and prevention of transistor saturation. This is due to the higher current density in the Schottky diode.

Reverse Current and Discharge Protection

Because of a Schottky diode's low forward voltage drop; less energy is wasted as heat, making them the most efficient choice for applications sensitive to efficiency. For instance, they are used in stand-alone ("off-grid") photovoltaic (PV) systems to prevent batteries from discharging through the solar panels at night, called "blocking diodes". They are also used in grid-connected systems with multiple strings connected in parallel, in order to prevent reverse current flowing from adjacent strings through shaded strings if the "bypass diodes" have failed.

Switched-Mode Power Supplies

Schottky diode are also used as rectifiers in switched-mode power supplies. The low forward voltage and fast recovery time leads to increased efficiency.

They can also be used in power supply "OR"ing circuits in products that have both an internal battery and a mains adapter input, or similar. However, the high reverse leakage current presents a problem in this case, as any high-impedance voltage sensing circuit (e.g., monitoring the battery voltage or detecting whether a mains adapter is present) will see the voltage from the other power source through the diode leakage.

Sample-and-Hold Circuits

Schottky diodes can be used in diode-bridge based sample and hold circuits. When compared to regular p-n junction based diode bridges, Schottky diodes can offer advantages. A forward-biased Schottky diode does not have any minority carrier charge storage. This allows them to switch more quickly than regular diodes, resulting in lower transition time from the sample to the hold step. The absence of minority carrier charge storage also results in a lower hold step or sampeling error, resulting in a more accurate sample at the output.

Designation

Commonly encountered schottky diodes include the 1N58xx series rectifiers, such as the 1N581x (1 ampere) and 1N582x (3 ampere) through-hole parts, and the SS1x (1 ampere) and SS3x (3 ampere) surface-mount parts. Schottky rectifiers are available in numerous surface-mount package styles.

SS14 schottky diode in a DO-214AC (SMA) package (surface mount version of 1N5819)

Small-signal schottky diodes such as the 1N5711, 1N6263, 1SS106, 1SS108, and the BAT41–43, 45–49 series are widely used in high-frequency applications as detectors, mixers and nonlinear elements, and have superseded germanium diodes. They are also suitable for electrostatic discharge (ESD) protection of sensitive devices such as III-V-semiconductor devices, laser diodes and, to a lesser extent, exposed lines of CMOS circuitry.

Schottky metal–semiconductor junctions are featured in the successors to the 7400 TTL family of logic devices, the 74S, 74LS and 74ALS series, where they are employed as Baker clamps in parallel with the collector-base junctions of the bipolar transistors to prevent their saturation, thereby greatly reducing their turn-off delays.

Alternatives

When less power dissipation is desired, a MOSFET and a control circuit can be used instead, in an operation mode known as active rectification.

A super diode consisting of a pn-diode or Schottky diode and an operational amplifier provides an almost perfect diode characteristic due to the effect of negative feedback, although its use is restricted to frequencies the operational amplifier used can handle.

Current–Voltage Characteristic

A semiconductor diode's behavior in a circuit is given by its current–voltage characteristic, or I–V graph (see graph below). The shape of the curve is determined by the transport of charge carriers through the so-called depletion layer or depletion region that exists at the p–n junction between differing semiconductors. When a p–n junction is first created, conduction-band (mobile) electrons from the N-doped region diffuse into the P-doped region where there is a large population of holes (vacant places for electrons) with which the electrons "recombine". When a mobile electron recombines with a hole, both hole and electron vanish, leaving behind an immobile positively charged donor (dopant) on the N side and negatively charged acceptor (dopant) on the P side. The region around the p–n junction becomes depleted of charge carriers and thus behaves as an insulator.

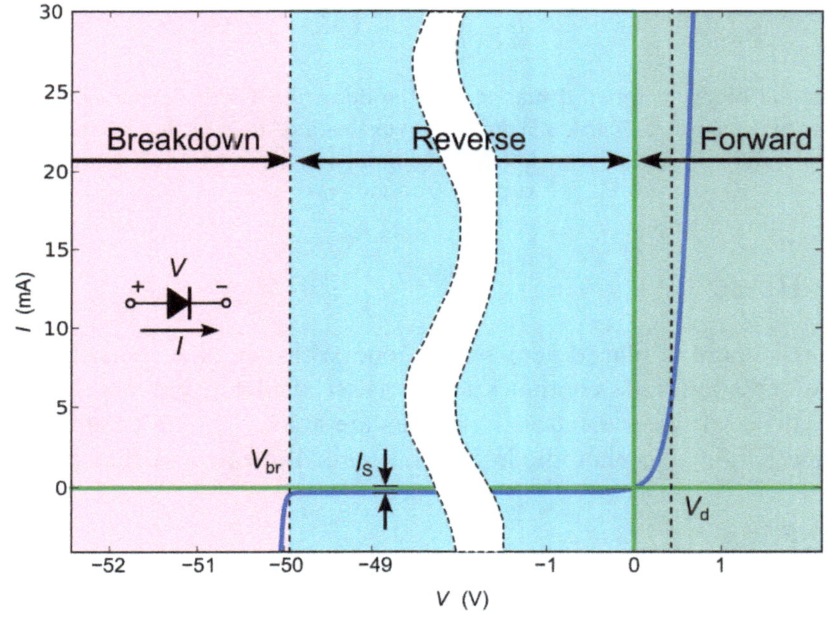

I–V (current vs. voltage) characteristics of a p–n junction diode

However, the width of the depletion region (called the depletion width) cannot grow without limit. For each electron–hole pair recombination made, a positively charged dopant ion is left behind in the N-doped region, and a negatively charged dopant ion is created in the P-doped region. As recombination proceeds and more ions are created, an increasing electric field develops through the depletion zone that acts to slow and then finally stop recombination. At this point, there is a "built-in" potential across the depletion zone.

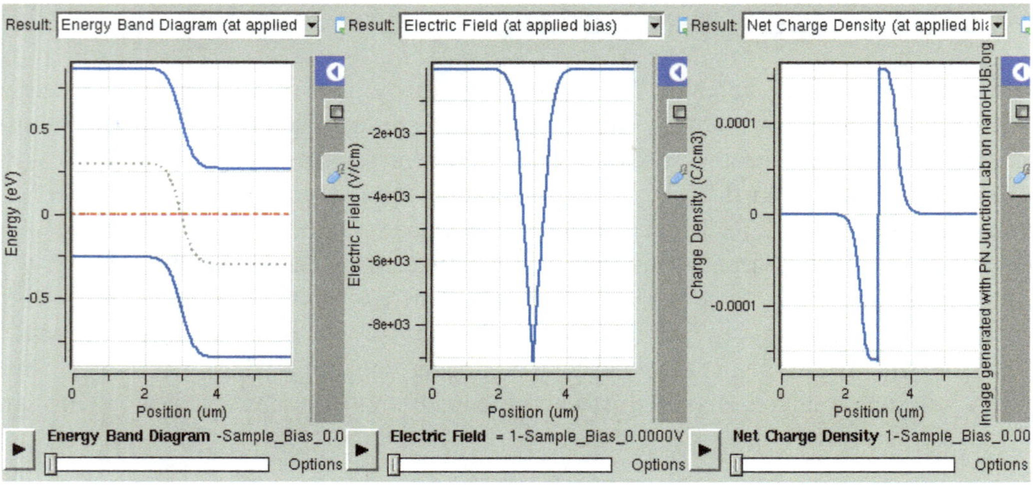

A PN junction diode in forward bias mode, the depletion width decreases. Both p and n junctions are doped at a 1e15/cm3 doping level, leading to built-in potential of ~0.59V. Observe the different Quasi Fermi levels for conduction band and valence band in n and p regions (red curves).

Reverse Bias

If an external voltage is placed across the diode with the same polarity as the built-in potential, the depletion zone continues to act as an insulator, preventing any significant electric current flow (unless electron–hole pairs are actively being created in the junction by, for instance, light; see photodiode). This is called the reverse bias phenomenon.

Forward Bias

However, if the polarity of the external voltage opposes the built-in potential, recombination can once again proceed, resulting in a substantial electric current through the p–n junction (i.e. substantial numbers of electrons and holes recombine at the junction). For silicon diodes, the built-in potential is approximately 0.7 V (0.3 V for germanium and 0.2 V for Schottky). Thus, if an external voltage greater than and opposite to the built-in voltage is applied, a current will flow and the diode is said to be "turned on" as it has been given an external forward bias. The diode is commonly said to have a forward "threshold" voltage, above which it conducts and below which conduction stops. However, this is only an approximation as the forward characteristic is according to the Shockley equation absolutely smooth (see graph below).

A diode's I–V characteristic can be approximated by four regions of operation:

- At very large reverse bias, beyond the peak inverse voltage or PIV, a process called reverse breakdown occurs that causes a large increase in current (i.e., a large number of electrons and holes are created at, and move away from the p–n junction) that usually damages the device permanently. The avalanche diode is deliberately designed for use in that manner. In the Zener diode, the concept of PIV is not applicable. A Zener diode contains a heavily doped p–n junction allowing electrons to tunnel from the valence band of the p-type material to the conduction band of the n-type material, such that the reverse voltage is "clamped" to a known value (called the Zener voltage), and avalanche does not occur. Both devices, however, do have a limit to the maximum current and power they can withstand in the clamped reverse-voltage region. Also, following the end of forward conduction in any diode, there is reverse current for a short time. The device does not attain its full blocking capability until the reverse current ceases.

- For a bias less than the PIV, the reverse current is very small. For a normal P–N rectifier diode, the reverse current through the device in the micro-ampere (μA) range is very low. However, this is temperature dependent, and at sufficiently high temperatures, a substantial amount of reverse current can be observed (mA or more).
- With a small forward bias, where only a small forward current is conducted, the current–voltage curve is exponential in accordance with the ideal diode equation. There is a definite forward voltage at which the diode starts to conduct significantly. This is called the knee voltage or cut-in voltage and is equal to the barrier potential of the p-n junction. This is a feature of the exponential curve, and appears sharper on a current scale more compressed than in the diagram shown here.
- At larger forward currents the current-voltage curve starts to be dominated by the ohmic resistance of the bulk semiconductor. The curve is no longer exponential, it is asymptotic to a straight line whose slope is the bulk resistance. This region is particularly important for power diodes. The diode can be modeled as an ideal diode in series with a fixed resistor.

In a small silicon diode operating at its rated currents, the voltage drop is about 0.6 to 0.7 volts. The value is different for other diode types—Schottky diodes can be rated as low as 0.2 V, germanium diodes 0.25 to 0.3 V, and red or blue light-emitting diodes (LEDs) can have values of 1.4 V and 4.0 V respectively.

At higher currents the forward voltage drop of the diode increases. A drop of 1 V to 1.5 V is typical at full rated current for power diodes.

Shockley Diode Equation

The Shockley ideal diode equation or the diode law (named after transistor co-inventor William Bradford Shockley) gives the I–V characteristic of an ideal diode in either forward or reverse bias (or no bias). The following equation is called the Shockley ideal diode equation when n, the ideality factor, is set equal to 1 :

$$I = I_S \left(e^{\frac{V_D}{nV_T}} - 1 \right)$$

where

I is the diode current,

I_S is the reverse bias saturation current (or scale current),

V_D is the voltage across the diode,

V_T is the thermal voltage, and

n is the ideality factor, also known as the quality factor or sometimes emission coefficient. The ideality factor n typically varies from 1 to 2 (though can in some cases be higher), depending on the fabrication process and semiconductor material and is set equal to 1 for the case of an "ideal" diode (thus the n is sometimes omitted). The ideality factor was added to account for imperfect junctions as observed in real transistors. The factor mainly accounts for carrier recombination as the charge carriers cross the depletion region.

The thermal voltage V_T is approximately 25.85 mV at 300 K, a temperature close to "room temperature" commonly used in device simulation software. At any temperature it is a known constant defined by:

$$V_T = \frac{kT}{q},$$

where k is the Boltzmann constant, T is the absolute temperature of the p–n junction, and q is the magnitude of charge of an electron (the elementary charge).

The reverse saturation current, I_S, is not constant for a given device, but varies with temperature; usually more significantly than V_T, so that V_D typically decreases as T increases.

The Shockley ideal diode equation or the diode law is derived with the assumption that the only processes giving rise to the current in the diode are drift (due to electrical field), diffusion, and thermal recombination–generation (R–G) (this equation is derived by setting n = 1 above). It also assumes that the R–G current in the depletion region is insignificant. This means that the Shockley ideal diode equation doesn't account for the processes involved in reverse breakdown and photon-assisted R–G. Additionally, it doesn't describe the "leveling off" of the I–V curve at high forward bias due to internal resistance. Introducing the ideality factor, n, accounts for recombination and generation of carriers.

Under reverse bias voltages the exponential in the diode equation is negligible, and the current is a constant (negative) reverse current value of $-I_S$. The reverse breakdown region is not modeled by the Shockley diode equation.

For even rather small forward bias voltages the exponential is very large, since the thermal voltage is very small in comparison. The subtracted '1' in the diode equation is then negligible and the forward diode current can be approximated by

$$I = I_S e^{\frac{V_D}{nV_T}}$$

The use of the diode equation in circuit problems is illustrated in the article on diode modeling.

Small-Signal Behavior

For circuit design, a small-signal model of the diode behavior often proves useful. A specific example of diode modeling is discussed in the article on small-signal circuits.

Reverse-Recovery Effect

Following the end of forward conduction in a p–n type diode, a reverse current can flow for a short time. The device does not attain its blocking capability until the mobile charge in the junction is depleted.

The effect can be significant when switching large currents very quickly. A certain amount of "reverse recovery time" t_r (on the order of tens of nanoseconds to a few microseconds) may be required to remove the reverse recovery charge Q_r from the diode. During this recovery time, the diode can actually conduct in the reverse direction. This might give rise to a large constant current in the reverse direction for a short time while the diode is reverse biased. The magnitude of such a reverse current is determined by the operating circuit (i.e., the series resistance) and the diode is said to be in the storage-phase. In certain real-world cases it is important to consider the losses that are incurred by this non-ideal diode effect. However, when the slew rate of the current is not so severe (e.g. Line frequency) the effect can be safely ignored. For most applications, the effect is also negligible for Schottky diodes.

The reverse current ceases abruptly when the stored charge is depleted; this abrupt stop is exploited in step recovery diodes for generation of extremely short pulses.

TYPES OF SEMICONDUCTOR DIODE

There are several types of p–n junction diodes, which emphasize either a different physical aspect of a diode often by geometric scaling, doping level, choosing the right electrodes, are just an application of a diode in a special circuit, or are really different devices like the Gunn and laser diode and the MOSFET:

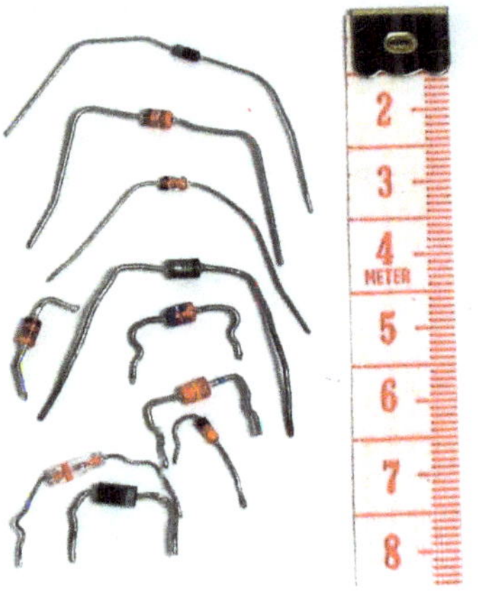

Several types of diodes. The scale is centimeters.

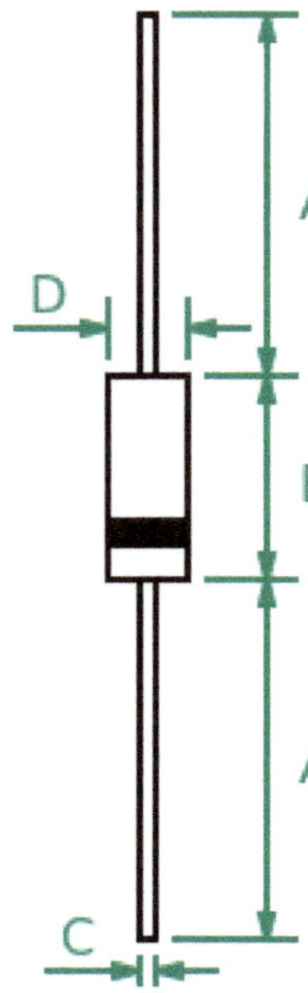

Typical datasheet drawing showing the dimensions of a DO-41 diode package

Normal (p–n) diodes, which operate as described above, are usually made of doped silicon or, more rarely, germanium. Before the development of silicon power rectifier diodes, cuprous oxide and later selenium was used. Their low efficiency required a much higher forward voltage to be applied (typically 1.4 to 1.7 V per "cell", with multiple cells stacked so as to increase the peak inverse voltage rating for application in high voltage rectifiers), and required a large heat sink (often an extension of the diode's metal substrate), much larger than the later silicon diode of the same current ratings would require. The vast majority of all diodes are the p–n diodes found in CMOS integrated circuits, which include two diodes per pin and many other internal diodes.

Avalanche Diodes

These are diodes that conduct in the reverse direction when the reverse bias voltage exceeds the breakdown voltage. These are electrically very similar to Zener diodes (and are often mistakenly called Zener diodes), but break down by a different mechanism: the avalanche effect. This occurs when the reverse electric field applied across the p–n junction causes a wave of ionization, reminiscent of an avalanche, leading to a large current. Avalanche diodes are designed to break down at a well-defined reverse voltage without being destroyed. The difference between the avalanche diode (which has a reverse breakdown above about 6.2 V) and the Zener is that the channel length of the former exceeds the mean free path of the electrons, resulting in many collisions between them on the way through the channel. The only practical difference between the two types is they have temperature coefficients of opposite polarities.

Cat's Whisker or Crystal Diodes

These are a type of point-contact diode. The cat's whisker diode consists of a thin or sharpened metal wire pressed against a semiconducting crystal, typically galena or a piece of coal. The wire forms the anode and the crystal forms the cathode. Cat's whisker diodes were also called crystal diodes and found application in the earliest radios called crystal radio receivers. Cat's whisker diodes are generally obsolete, but may be available from a few manufacturers.

Constant current Diodes

These are actually JFETs with the gate shorted to the source, and function like a two-terminal current-limiting analog to the voltage-limiting Zener diode. They allow a current through them to rise to a certain value, and then level off at a specific value. Also called CLDs, constant-current diodes, diode-connected transistors, or current-regulating diodes.

Esaki or Tunnel Diodes

These have a region of operation showing negative resistance caused by quantum tunneling, allowing amplification of signals and very simple bistable circuits. Because of the high carrier concentration, tunnel diodes are very fast, may be used at low (mK) temperatures, high magnetic fields, and in high radiation environments. Because of these properties, they are often used in spacecraft.

Gunn Diodes

These are similar to tunnel diodes in that they are made of materials such as GaAs or InP that exhibit a region of negative differential resistance. With appropriate biasing, dipole domains form and travel across the diode, allowing high frequency microwave oscillators to be built.

Light-Emitting Diodes (LEDs)

In a diode formed from a direct band-gap semiconductor, such as gallium arsenide, charge carriers that cross the junction emit photons when they recombine with the majority carrier on the other side. Depending on the material, wavelengths (or colors) from the infrared to the near ultraviolet may be produced. The forward potential of these diodes depends on the wavelength of the emitted photons: 2.1 V corresponds to red, 4.0 V to violet. The first LEDs were red and yellow, and higher-frequency diodes have been developed over time. All LEDs produce incoherent, narrow-spectrum light; "white" LEDs are actually combinations of three LEDs of a different color, or a blue LED with a yellow scintillator coating. LEDs can also be used as low-efficiency photodiodes in signal applications. An LED may be paired with a photodiode or phototransistor in the same package, to form an opto-isolator.

Laser Diodes

When an LED-like structure is contained in a resonant cavity formed by polishing the parallel end faces, a laser can be formed. Laser diodes are commonly used in optical storage devices and for high speed optical communication.

Thermal Diodes

This term is used both for conventional p–n diodes used to monitor temperature because of their varying forward voltage with temperature, and for Peltier heat pumps for thermoelectric heating and cooling. Peltier heat pumps may be made from semiconductor, though they do not have any rectifying junctions, they use the differing behavior of charge carriers in N and P type semiconductor to move heat.

Perun's Diodes

This is a special type of voltage-surge protection diode. It is characterized by the symmetrical voltage-current characteristic, similar to DIAC. It has much faster response time however, that's why it is used in demanding applications.

Photodiodes

All semiconductors are subject to optical charge carrier generation. This is typically an undesired effect, so most semiconductors are packaged in light blocking material. Photodiodes are intended to sense light(photodetector), so they are packaged in materials that allow light to pass, and are usually PIN (the kind of diode most sensitive to light). A photodiode can be used in solar cells, in photometry, or in optical communications. Multiple photodiodes may be packaged in a single device, either as a linear array or as a two-dimensional array. These arrays should not be confused with charge-coupled devices.

PIN Diodes

A PIN diode has a central un-doped, or intrinsic, layer, forming a p-type/intrinsic/n-type structure. They are used as radio frequency switches and attenuators. They are also used as large-volume, ionizing-radiation detectors and as photodetectors. PIN diodes are also used in power electronics, as their central layer can withstand high voltages. Furthermore, the PIN structure can be found in many power semiconductor devices, such as IGBTs, power MOSFETs, and thyristors.

Schottky Diodes

Schottky diodes are constructed from a metal to semiconductor contact. They have a lower forward voltage drop than p–n junction diodes. Their forward voltage drop at forward currents of about 1 mA is in the range 0.15 V to 0.45 V, which makes them useful in voltage clamping applications and prevention of transistor saturation. They can also be used as low loss rectifiers, although their reverse leakage current is in general higher than that of other diodes. Schottky diodes are majority carrier devices and so do not suffer from minority carrier storage problems that slow down many other diodes—so they have a faster reverse recovery than p–n junction diodes. They also tend to have much lower junction capacitance than p–n diodes, which provides for high switching speeds and their use in high-speed circuitry and RF devices such as switched-mode power supply, mixers, and detectors.

Super Barrier Diodes

Super barrier diodes are rectifier diodes that incorporate the low forward voltage drop of the Schottky diode with the surge-handling capability and low reverse leakage current of a normal p–n junction diode.

Gold-Doped Diodes

As a dopant, gold (or platinum) acts as recombination centers, which helps a fast recombination of minority carriers. This allows the diode to operate at signal frequencies, at the expense of a higher forward voltage drop. Gold-doped diodes are faster than other p–n diodes (but not as fast as Schottky diodes). They also have less reverse-current leakage than Schottky diodes (but not as good as other p–n diodes). A typical example is the 1N914.

Snap-Off or Step Recovery Diodes

The term step recovery relates to the form of the reverse recovery characteristic of these devices. After a forward current has been passing in an SRD and the current is interrupted or reversed, the reverse conduction will cease very abruptly (as in a step waveform). SRDs can, therefore, provide very fast voltage transitions by the very sudden disappearance of the charge carriers.

Stabistors or Forward Reference Diodes

The term stabistor refers to a special type of diodes featuring extremely stable forward voltage characteristics. These devices are specially designed for low-voltage stabilization applications requiring a guaranteed voltage over a wide current range and highly stable over temperature.

Transient Voltage Suppression Diode (TVS)

These are avalanche diodes designed specifically to protect other semiconductor devices from high-voltage transients. Their p–n junctions have a much larger cross-sectional area than those of a normal diode, allowing them to conduct large currents to ground without sustaining damage.

Varicap or Varactor Diodes

These are used as voltage-controlled capacitors. These are important in PLL (phase-locked loop) and FLL (frequency-locked loop) circuits, allowing tuning circuits, such as those in television receivers, to lock quickly on to the frequency. They also enabled tunable oscillators in early discrete tuning of radios, where a cheap and stable, but fixed-frequency, crystal oscillator provided the reference frequency for a voltage-controlled oscillator.

Zener Diodes

These can be made to conduct in reverse bias (backward), and are correctly termed reverse breakdown diodes. This effect, called Zener breakdown, occurs at a precisely defined voltage, allowing the diode to be used as a precision voltage reference. The term Zener diode is colloquially applied to several types of breakdown diodes, but strictly speaking Zener diodes have a breakdown voltage of below 5 volts, whilst avalanche diodes are used for breakdown voltages above that value. In practical voltage reference circuits, Zener and switching diodes are connected in series and opposite directions to balance the temperature coefficient response of the diodes to near-zero. Some devices labeled as high-voltage Zener diodes are actually avalanche diodes. Two (equivalent) Zeners in series and in reverse order, in the same package, constitute a transient absorber (or Transorb, a registered trademark).

Other uses for semiconductor diodes include the sensing of temperature, and computing analog logarithms.

Numbering and Coding Schemes

There are a number of common, standard and manufacturer-driven numbering and coding schemes for diodes; the two most common being the EIA/JEDEC standard and the European Pro Electron standard:

EIA/JEDEC

The standardized 1N-series numbering EIA370 system was introduced in the US by EIA/JEDEC (Joint Electron Device Engineering Council) about 1960. Most diodes have a 1-prefix designation (e.g., 1N4003). Among the most popular in this series were: 1N34A/1N270 (germanium signal), 1N914/1N4148 (silicon signal), 1N400x (silicon 1A power rectifier), and 1N580x (silicon 3A power rectifier).

JIS

The JIS semiconductor designation system has all semiconductor diode designations starting with “1S”.

Pro Electron

The European Pro Electron coding system for active components was introduced in 1966 and comprises two letters followed by the part code. The first letter represents the semiconductor material used for the component (A = germanium and B = silicon) and the second letter represents the general function of the part (for diodes, A = low-power/ signal, B = variable capacitance, X = multiplier, Y = rectifier and Z = voltage reference); for example:

- AA-series germanium low-power/signal diodes (e.g., AA119)
- BA-series silicon low-power/signal diodes (e.g., BAT18 silicon RF switching diode)
- BY-series silicon rectifier diodes (e.g., BY127 1250V, 1A rectifier diode)
- BZ-series silicon Zener diodes (e.g., BZY88C4V7 4.7V Zener diode)

Other common numbering / coding systems (generally manufacturer-driven) include:

- GD-series germanium diodes (e.g., GD9) – this is a very old coding system
- OA-series germanium diodes (e.g., OA47) – a coding sequence developed by Mullard, a UK company

As well as these common codes, many manufacturers or organisations have their own systems too – for example:

- HP diode 1901-0044 = JEDEC 1N4148
- UK military diode CV448 = Mullard type OA81 = GEC type GEX23

Related Devices

- Rectifier
- Transistor
- Thyristor or silicon controlled rectifier (SCR)
- TRIAC
- DIAC
- Varistor

In optics, an equivalent device for the diode but with laser light would be the Optical isolator, also known as an Optical Diode, that allows light to only pass in one direction. It uses a Faraday rotator as the main component.

APPLICATIONS

Radio Demodulation

The first use for the diode was the demodulation of amplitude modulated (AM) radio broadcasts. The history of this discovery is treated in depth in the radio article. In summary, an AM signal consists of alternating positive and negative peaks of a radio carrier wave, whose amplitude or envelope is proportional to the original audio signal. The diode (originally a crystal diode) rectifies the AM radio frequency signal, leaving only the positive peaks of the carrier wave. The audio is then extracted from the rectified carrier wave using a simple filter and fed into an audio amplifier or transducer, which generates sound waves.

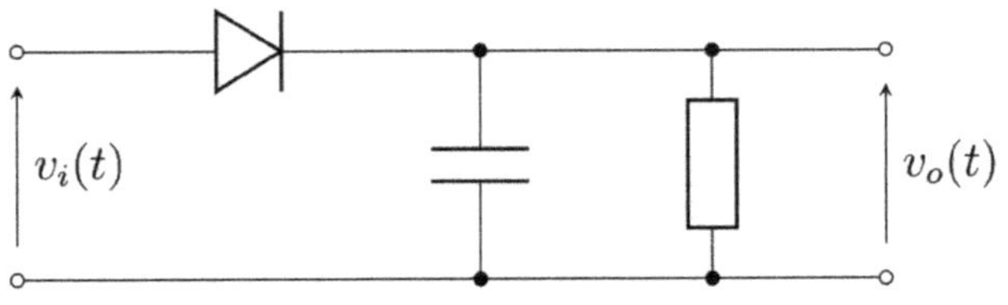

A simple envelope demodulator circuit.

Power Conversion

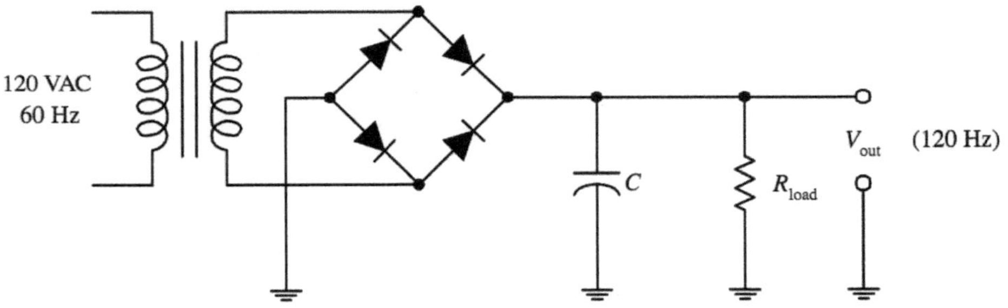

Schematic of basic AC-to-DC power supply

Rectifiers are constructed from diodes, where they are used to convert alternating current (AC) electricity into direct current (DC). Automotive alternators are a common example, where the diode, which rectifies the AC into DC, provides better performance than the commutator or earlier, dynamo. Similarly, diodes are also used in Cockcroft–Walton voltage multipliers to convert AC into higher DC voltages.

Over-Voltage Protection

Diodes are frequently used to conduct damaging high voltages away from sensitive electronic devices. They are usually reverse-biased (non-conducting) under normal circumstances. When the voltage rises above the normal range, the diodes become forward-biased (conducting). For example, diodes are used in (stepper motor and H-bridge) motor controller and relay circuits to de-energize coils rapidly without the damaging voltage spikes that would otherwise occur. (A diode used in such an application is called a flyback diode).

Many integrated circuits also incorporate diodes on the connection pins to prevent external voltages from damaging their sensitive transistors. Specialized diodes are used to protect from over-voltages at higher power.

Logic Gates

Diodes can be combined with other components to construct AND and OR logic gates. This is referred to as diode logic.

Ionizing Radiation Detectors

In addition to light, mentioned above, semiconductor diodes are sensitive to more energetic radiation. In electronics, cosmic rays and other sources of ionizing radiation cause noise pulses and single and multiple bit errors. This effect is sometimes exploited by particle detectors to detect radiation. A single particle of radiation, with thousands or millions of electron volts of energy, generates many charge carrier pairs, as its energy is deposited in the semiconductor material. If the depletion layer is large enough to catch the whole shower or to stop a heavy particle, a fairly accurate measurement of the particle's energy can be made, simply by measuring the charge conducted and without the complexity of a magnetic spectrometer, etc. These semiconductor radiation detectors need efficient and uniform charge collection and low leakage current. They are often cooled by liquid nitrogen. For longer-range (about a centimeter) particles, they need a very large depletion depth and large area. For short-range particles, they need any contact or un-depleted semiconductor on at least one surface to be very thin. The back-bias voltages are near breakdown (around a thousand volts per centimeter). Germanium and silicon are common materials. Some of these detectors sense position as well as energy. They have a finite life, especially when detecting heavy particles, because of radiation damage. Silicon and germanium are quite different in their ability to convert gamma rays to electron showers.

Semiconductor detectors for high-energy particles are used in large numbers. Because of energy loss fluctuations, accurate measurement of the energy deposited is of less use.

Temperature Measurements

A diode can be used as a temperature measuring device, since the forward voltage drop across the diode depends on temperature, as in a silicon bandgap temperature sensor. From the Shockley ideal diode equation given above, it might appear that the voltage has a positive temperature coefficient (at a constant current), but usually the variation of the reverse saturation current term is more significant than the variation in the thermal voltage term. Most diodes therefore have a negative temperature coefficient, typically −2 mV/°C for silicon diodes. The temperature coefficient is approximately constant for temperatures above about 20 kelvins. Some graphs are given for 1N400x series, and CY7 cryogenic temperature sensor.

Current Steering

Diodes will prevent currents in unintended directions. To supply power to an electrical circuit during a power failure, the circuit can draw current from a battery. An uninterruptible power supply may use diodes in this way to ensure that current is only drawn from the battery when necessary. Likewise, small boats typically have two circuits each with their own battery/batteries: one used for engine starting; one used for domestics. Normally, both are charged from a single alternator, and a heavy-duty split-charge diode is used to prevent the higher-charge battery (typically the engine battery) from discharging through the lower-charge battery when the alternator is not running.

Diodes are also used in electronic musical keyboards. To reduce the amount of wiring needed in electronic musical keyboards, these instruments often use keyboard matrix circuits. The keyboard controller scans the rows and columns to determine which note the player has pressed. The problem with matrix circuits is that, when several notes are pressed at once, the current can flow backwards through the circuit and trigger "phantom keys" that cause "ghost" notes to play. To avoid triggering unwanted notes, most keyboard matrix circuits have diodes soldered with the switch under each key of the musical keyboard. The same principle is also used for the switch matrix in solid-state pinball machines.

Waveform Clipper

Diodes can be used to limit the positive or negative excursion of a signal to a prescribed voltage.

Clamper

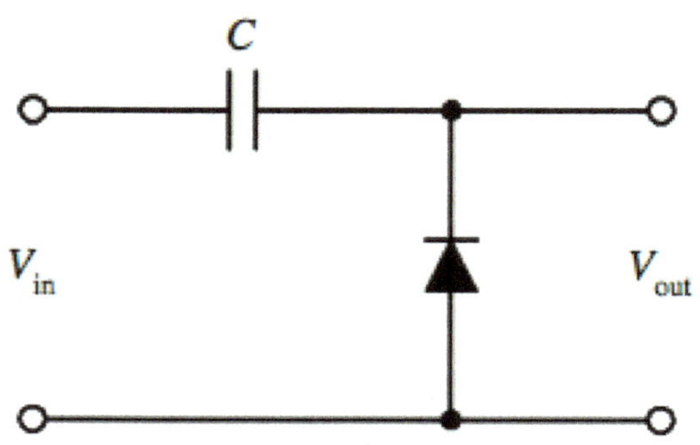

This simple diode clamp will clamp the negative peaks of the incoming waveform to the common rail voltage

A diode clamp circuit can take a periodic alternating current signal that oscillates between positive and negative values, and vertically displace it such that either the positive, or the negative peaks occur at a prescribed level. The clamper does not restrict the peak-to-peak excursion of the signal, it moves the whole signal up or down so as to place the peaks at the reference level.

REFERENCES

1. Tooley, Mike (2012). Electronic Circuits: Fundamentals and Applications, 3rd Ed. Routlege. p. 81. ISBN 1-136-40731-6.
2. Lowe, Doug (2013). "Electronics Components: Diodes". Electronics All-In-One Desk Reference For Dummies. John Wiley & Sons. Retrieved January 4, 2013.
3. Crecraft, David; Stephen Gergely (2002). Analog Electronics: Circuits, Systems and Signal Processing. Butterworth-Heinemann. p. 110. ISBN 0-7506-5095-8.
4. Horowitz, Paul; Winfield Hill (1989). The Art of Electronics, 2nd Ed. London: Cambridge University Press. p. 44. ISBN 0-521-37095-7.
5. "Physical Explanation – General Semiconductors". 2010-05-25. Retrieved 2010-08-06.
6. "The Constituents of Semiconductor Components". 2010-05-25. Retrieved 2010-08-06.
7. Guthrie, Frederick (October 1873) "On a relation between heat and static electricity," The London, Edinburgh and Dublin Philosophical Magazine and Journal of Science, 4th series, 46 : 257–266.
8. 1928 Nobel Lecture: Owen W. Richardson, "Thermionic phenomena and the laws which govern them", December 12, 1929
9. Edison, Thomas A. "Electrical Meter" U.S. Patent 307,030 Issue date: Oct 21, 1884
10. "Road to the Transistor". Jmargolin.com. Retrieved 2008-09-22.
11. Braun, Ferdinand (1874) "Ueber die Stromleitung durch Schwefelmetalle" (On current conduction in metal sulphides), Annalen der Physik und Chemie, 153 : 556–563.
12. Karl Ferdinand Braun. chem.ch.huji.ac.il
13. "Diode". Encyclobeamia.solarbotics.net.
14. Sarkar, Tapan K. (2006). History of wireless. USA: John Wiley and Sons. pp. 94, 291–308. ISBN 0-471-71814-9.
15. Pickard, Greenleaf Whittier "Means for receiving intelligence communicated by electric waves" U.S. Patent 836,531 Issued: August 30, 1906
16. Choi, Charles Q. (4 April 2016) World's Smallest Diode Is Made of DNA. IEEE Spectrum
17. Emmino, Nicolette (5 April 2016) The World's Smallest Diode Has Been Developed. IHS Electronics360
18. Diode reverse recovery in a boost converter. ECEN5817. ecee.colorado.edu

19. Elhami Khorasani, A.; Griswold, M.; Alford, T. L. (2014). "Gate-Controlled Reverse Recovery for Characterization of LDMOS Body Diode". IEEE Electron Device Letters. 35 (11): 1079. Bibcode:2014IEDL...35.1079E. doi:10.1109/LED.2014.2353301.
20. Inclusion of Switching Loss in the Averaged Equivalent Circuit Model. ECEN5797. ecee.colorado.edu
21. Current regulator diodes. Digikey.com (2009-05-27). Retrieved 2013-12-19.
22. Jonscher, A. K. (1961). "The physics of the tunnel diode". British Journal of Applied Physics. 12 (12): 654. Bibcode:1961BJAP...12..654J. doi:10.1088/0508-3443/12/12/304.
23. Dowdey, J. E.; Travis, C. M. (1964). "An Analysis of Steady-State Nuclear Radiation Damage of Tunnel Diodes". IEEE Transactions on Nuclear Science. 11 (5): 55. Bibcode:1964ITNS...11...55D. doi:10.1109/TNS2.1964.4315475.
24. Classification of components. Digikey.com (2009-05-27). Retrieved 2013-12-19.
25. "Component Construction". 2010-05-25. Retrieved 2010-08-06.
26. Component Construction. Digikey.com (2009-05-27). Retrieved 2013-12-19.
27. "Physics and Technology". 2010-05-25. Retrieved 2010-08-06.
28. Fast Recovery Epitaxial Diodes (FRED) Characteristics – Applications – Examples. (PDF). Retrieved 2013-12-19.
29. Sze, S. M. (1998) Modern Semiconductor Device Physics, Wiley Interscience, ISBN 0-471-15237-4
30. Protecting Low Current Loads in Harsh Electrical Environments. Digikey.com (2009-05-27). Retrieved 2013-12-19.
31. "About JEDEC". Jedec.org. Retrieved 2008-09-22.
32. "Introduction dates of common transistors and diodes?". EDAboard.com. 2010-06-10. Archived from the original on October 11, 2007. Retrieved 2010-08-06.
33. I.D.E.A. "Transistor Museum Construction Projects Point Contact Germanium Western Electric Vintage Historic Semiconductors Photos Alloy Junction Oral History". Semiconductormuseum.com. Retrieved 2008-09-22.
34. 1N400x Diode Family Forward Voltage. Cliftonlaboratories.com. Retrieved 2013-12-19.
35. Cryogenic Temperature Sensors. omega.com
36. John Ambrose Fleming (1919). The Principles of Electric Wave Telegraphy and Telephony. London: Longmans, Green. p. 550.
37. John Sparkes (1994). Semiconductor Devices (2nd ed.). CRC Press. p. 78. ISBN 0-7487-7382-7.
38. Naturally, this voltage depends upon the selected current level. This voltage for the p–n diode is taken variously as 0.7 V and 0.5 V; see AS Sedra and KF Smith (1998). "Chapter 3: Diodes". Microelectronic circuits (4th ed.). Oxford University Press. p. 134 & Figure 3.8. ISBN 0-19-511663-1..

39. Andrei Grebennikov (2011). "§2.1.1: Diodes: Operational principle". RF and Microwave Transmitter Design. J Wiley & Sons. p. 59. ISBN 0-470-52099-X.

40. Narain Arora (2007). Mosfet modeling for VLSI simulation: theory and practice. World Scientific. p. 539. ISBN 981-256-862-X. Jean-Pierre Colinge, Cynthia A. Colinge (2002). Physics of semiconductor devices (2nd ed.). Springer. p. 149. ISBN 1-4020-7018-7.

41. The varactor is a p–n diode operated in reverse bias. See, for example, V.S.Bagad (2009). "§5.8.1 Varactor diode: Working principle". Microwave and Radar Engineering (2nd ed.). Technical Publications Pune. ISBN 81-8431-121-4.

CHAPTER 2

Transistors

Chapter Contents

Transistors are tiny switches that can be triggered by electric signals. They are the basic building blocks of microchips, and roughly define the difference between electric and electronic devices. They permeate so many facets of our daily lives, in everything from milk cartons to laptops, illustrating just how useful they are.

A traditional mechanical switch either enables or disables the flow of electricity by physically connecting (or disconnecting) two ends of wire. In a transistor, a signal tells the device to either conduct or insulate, thereby enabling or disabling the flow of electricity. This property of acting like an insulator in some circumstances and like a conductor in others is unique to a special class of materials known as "semiconductors."

INTRODUCTION

A transistor is a semiconductor device used to amplify or switch electronic signals and electrical power. It is composed of semiconductor material usually with at least three terminals for connection to an external circuit. A voltage or current applied to one pair of the transistor's terminals controls the current through another pair of terminals. Because the controlled (output) power can be higher than the controlling (input) power, a transistor can amplify a signal. Today, some transistors are packaged individually, but many more are found embedded in integrated circuits.

The transistor is the fundamental building block of modern electronic devices, and is ubiquitous in modern electronic systems. Julius Lilienfeld patented a field-effect transistor in 1926 but it was not possible to actually construct a working device at that time. The first practically implemented device was a point-contact transistor invented in 1947 by American physicists John Bardeen, Walter Brattain, and William Shockley. The transistor revolutionized the field of electronics, and paved the way for smaller and cheaper radios, calculators, and computers, among other things. The transistor is on the list of IEEE milestones in electronics, and Bardeen, Brattain, and Shockley shared the 1956 Nobel Prize in Physics for their achievement.

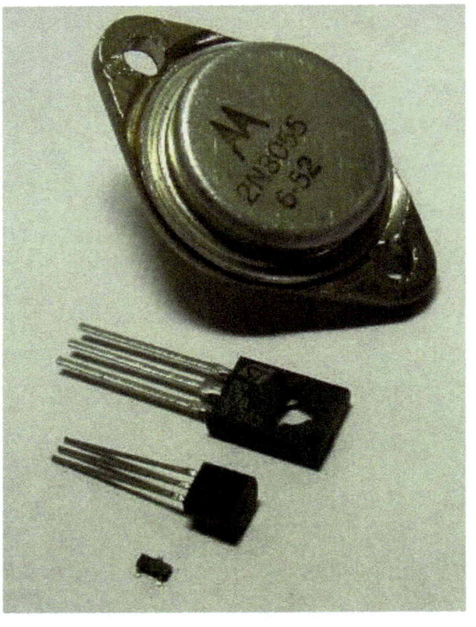

Assorted discrete transistors. Packages in order from top to bottom: TO-3, TO-126, TO-92, SOT-23.

HISTORY

A transistor is a semiconductor device with at least three terminals for connection to an electric circuit. The vacuum-tube triode, also called a (thermionic) valve, was the transistor's precursor, introduced in 1907.

Invention of the Transistor

The first patent for the field-effect transistor principle was filed in Canada by Austrian-Hungarian physicist Julius Edgar Lilienfeld on October 22, 1925, but Lilienfeld published no research articles about his devices, and his work was ignored by industry. In 1934 German physicist Dr. Oskar Heil patented another field-effect transistor. There is no direct evidence that these devices were built, but later work in the 1990s show that one of Lilienfeld's designs worked as described and gave substantial gain. Legal papers from the Bell Labs patent show that William Shockley and a co-worker at Bell Labs, Gerald Pearson, had built operational versions from Lilienfeld's patents, yet they never referenced this work in any of their later research papers or historical articles.

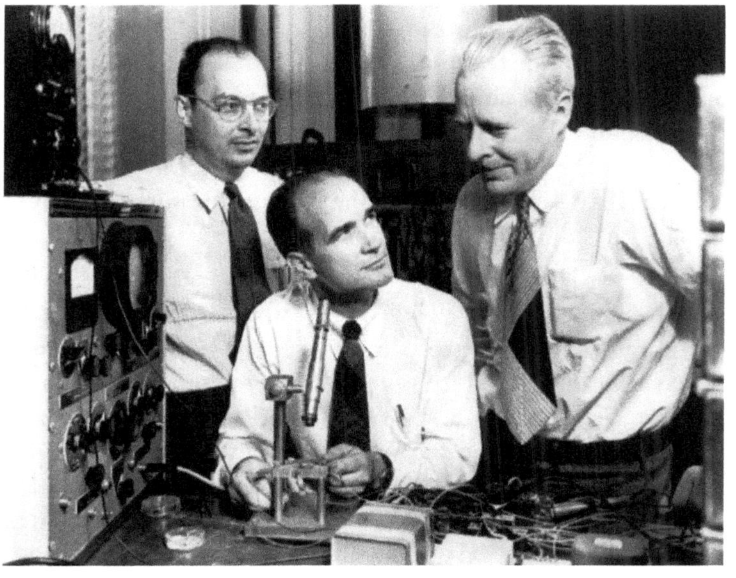

John Bardeen, William Shockley and Walter Brattain at Bell Labs, 1948.

The Bell Labs work on the transistor emerged from war-time efforts to produce extremely pure germanium "crystal" mixer diodes, used in radar units as a frequency mixer element in microwave radar receivers. A parallel project on germanium diodes at Purdue University succeeded in producing the good-quality germanium semiconducting crystals that were used at Bell Labs. Early tube-based circuits did not switch fast enough for this role, leading the Bell team to use solid state diodes instead.

After the war, Shockley decided to attempt the building of a triode-like semiconductor device. He secured funding and lab space, and went to work on the problem with Bardeen and Brattain. John Bardeen eventually developed a new branch of quantum mechanics known as surface physics to account for the "odd" behavior they saw, and Bardeen and Walter Brattain eventually succeeded in building a working device.

The key to the development of the transistor was the further understanding of the process of the electron mobility in a semiconductor. It was realized that if there was some way to control the flow of the electrons from the emitter to the collector of this newly discovered diode (discovered 1874; patented 1906), one could build an amplifier. For instance, if one placed contacts on either side of a single type of crystal the current would not flow through it. However, if a third contact could then "inject" electrons or holes into the material, the current would flow.

Actually doing this appeared to be very difficult. If the crystal were of any reasonable size, the number of electrons (or holes) required to be injected would have to be very large -- making it less useful than an amplifier because it would require a large injection current to start with. That said, the whole idea of the crystal diode was that the crystal itself could provide the electrons over a very small distance, the depletion region. The key appeared to be to place the input and output contacts very close together on the surface of the crystal on either side of this region.

Brattain started working on building such a device, and tantalizing hints of amplification continued to appear as the team worked on the problem. Sometimes the system would work but then stop working unexpectedly. In one instance a non-working system started working when placed in water. The electrons in any one piece of the crystal would migrate about due to nearby charges. Electrons in the emitters, or the "holes" in the collectors, would cluster at the surface of the crystal where they could find their opposite charge "floating around" in the air (or water). Yet they could be pushed away from the surface with the application of a small amount of charge from any other location on the crystal. Instead of needing a large supply of injected electrons, a very small number in the right place on the crystal would accomplish the same thing.

Their understanding solved the problem of needing a very small control area to some degree. Instead of needing two separate semiconductors connected by a common, but tiny, region, a single larger surface would serve. The emitter and collector leads would both be placed very close together on the top, with the control lead placed on the base of the crystal. When current was applied to the "base" lead, the electrons or holes would be pushed out, across the block of semiconductor, and collect on the far surface. As long as the emitter and collector were very close together, this should allow enough electrons or holes between them to allow conduction to start.

An early witness of the phenomenon was Ralph Bray, a young graduate student. He joined the germanium effort at Purdue University in November 1943 and was given the tricky task of measuring the spreading resistance at the metal-semiconductor contact. Bray found a great many anomalies, such as internal high-resistivity barriers in some samples of germanium. The most curious phenomenon was the exceptionally low resistance observed when voltage pulses were applied. This effect remained a mystery because nobody realized, until 1948, that Bray had observed minority carrier injection - the effect that was identified by William Shockley at Bell Labs and made the transistor a reality.

Bray wrote: "That was the one aspect that we missed, but even had we understood the idea of minority carrier injection...we would have said, 'Oh, this explains our effects.' We might not necessarily have gone ahead and said, 'Let's start making transistors,' open up a factory and sell them... At that time the important device was the high back voltage rectifier".

The First Transistor

The Bell team made many attempts to build such a system with various tools, but generally failed. Setups where the contacts were close enough were invariably as fragile as the original cat's whisker detectors had been, and would work briefly, if at all. Eventually they had a practical breakthrough. A piece of gold foil was glued to the edge of a plastic wedge, and then the foil was sliced with a razor at the tip of the triangle. The result was two very closely spaced contacts of gold. When the plastic was pushed down onto the surface of a crystal and voltage applied to the other side (on the base of the crystal), current started to flow from one contact to the other as the base voltage pushed the electrons away from the base towards the other side near the contacts. The point-contact transistor had been invented.

On 15 December 1947, "When the points were, very close together got voltage amp about 2 but not power amp. This voltage amplification was independent of frequency 10 to 10,000 cycles".

A stylized replica of the first transistor

On 16 December 1947, "Using this double point contact, contact was made to a germanium surface that had been anodized to 90 volts, electrolyte washed off in H_2O and then had some gold spots evaporated on it. The gold contacts were pressed down on the bare surface. Both gold contacts to the surface rectified nicely... The separation between points was about $4x10^{-3}$ cm. One point was used as a grid and the other point as a plate. The bias (D.C.) on the grid had to be positive to get amplification... power gain 1.3 voltage gain 15 on a plate bias of about 15 volts".

Brattain and H. R. Moore made a demonstration to several of their colleagues and managers at Bell Labs on the afternoon of 23 December 1947, often given as the birth date of the transistor. The "PNP point-contact germanium transistor" operated as a speech amplifier with a power gain of 18 in that trial. In 1956 John Bardeen, Walter Houser Brattain, and William Bradford Shockley were honored with the Nobel Prize in Physics "for their researches on semiconductors and their discovery of the transistor effect".

Twelve people are mentioned as directly involved in the invention of the transistor in the Bell Laboratory.

At the same time some European scientists were led by the idea of solid-state amplifiers. In August 1948 German physicists Herbert F. Mataré (1912–2011) and Heinrich Welker (1912–1981), working at Compagnie des Freins et Signaux Westinghouse in Aulnay-sous-Bois, France applied for a patent on an amplifier based on the minority carrier injection process which they called the "transistron". Since Bell Labs did not make a public announcement of the transistor until June 1948, the transistron was considered to be independently developed. Mataré had first observed transconductance effects during the manufacture of silicon diodes for German radar equipment during WWII. Transistrons were commercially manufactured for the French telephone company and military, and in 1953 a solid-state radio receiver with four transistrons was demonstrated at the Düsseldorf Radio Fair.

Origin of the term

Bell Telephone Laboratories needed a generic name for the new invention: "Semiconductor Triode", "Surface States Triode", "Crystal Triode", "Solid Triode" and "Iotatron" were all considered, but "Transistor," coined by John R. Pierce, was the clear winner of an internal ballot (owing in part to the affinity that Bell engineers had developed for the suffix "-istor").The rationale for the name is described in the following extract from the company's Technical Memorandum calling for votes:

Transistor. This is an abbreviated combination of the words "transconductance" or "transfer", and "varistor". The device logically belongs in the varistor family, and has the transconductance or transfer impedance of a device having gain, so that this combination is descriptive.

—Bell Telephone Laboratories — Technical Memorandum (May 28, 1948)

Pierce recalled the naming somewhat differently:

The way I provided the name, was to think of what the device did. And at that time, it was supposed to be the dual of the vacuum tube. The vacuum tube had transconductance, so the transistor would have 'transresistance.' And the name should fit in with the names of other devices, such as varistor and thermistor. And. . . I suggested the name 'transistor.'

—John R. Pierce, interviewed for PBS show "Transistorized!"

The Nobel Foundation states that the term is a combination of the words "transfer" and "resistor".

Production and Commercialization

The world's first commercial transistor production line was at the Western Electric plant on Union Boulevard in Allentown, Pennsylvania. Production began on Oct. 1, 1951 with the point contact germanium transistor.

By 1953, the transistor was being used in some products, such as hearing aids and telephone exchanges, but there were still significant issues preventing its broader application, such as sensitivity to moisture and the fragility of the wires attached to germanium crystals. Donald G. Fink, Philco's director of research, summarized the status of the transistor's commercial potential with an analogy: "Is it a pimpled adolescent, now awkward, but promising future vigor? Or has it arrived at maturity, full of languor, surrounded by disappointments?"

Transistor Radios

Prototypes of all-transistor AM radio receivers were demonstrated, but were really only laboratory curiosities. However, in 1950 Shockley developed a radically different type of solid-state amplifier which became known as the bipolar junction "transistor". Although it works on a completely different principle to the point-contact "transistor", this is the device which is most commonly referred to as simply a "transistor" today. Morgan Sparks made the bipolar junction transistor into a practical device. These were also licensed to a number of other electronics companies, including Texas Instruments, who produced a limited run of transistor radios as a sales tool. Early transistors were chemically unstable and only suitable for low-power, low-frequency applications, but as transistor design developed, these problems were slowly overcome.

The Regency TR-1 which used Texas Instruments' NPN transistors was the world's first commercially produced transistor radio.

There are numerous claimants to the title of the first company to produce practical transistor radios. Texas Instruments had demonstrated all-transistor AM radios as early as 1952, but their performance was well below that of equivalent battery tube models. A workable all-transistor radio was demonstrated in August 1953 at the Düsseldorf Radio Fair by the German firm Intermetall. It was built with four of Intermetall's hand-made transistors, based upon the 1948 invention of Herbert Mataré and Heinrich Welker.

However, as with the early Texas units (and others) only prototypes were ever built; it was never put into commercial production.

The production of the first commercially successful transistor radio is often incorrectly attributed to Sony (originally Tokyo Tsushin Kogyo). However the Regency TR-1, made by the Regency Division of I.D.E.A. (Industrial Development Engineering Associates) of Indianapolis, Indiana, was the first practical transistor radio made in any significant numbers. The TR-1 was announced on October 18, 1954 and put on sale in November 1954 for $49.95 (the equivalent of about $361 in year-2005 dollars) and sold about 150,000 units.

The TR-1 used four Texas NPN transistors and had to be powered by a 22.5-volt battery, since the only way to get adequate radio frequency performance out of early transistors was to run them close to their collector-to-emitter breakdown voltage. This made the TR-1 very expensive to run, and it was far more popular for its novelty or status value that its actual performance, rather in the fashion of the first MP3 players.

Still, aside from its indifferent performance, the TR-1 was a very advanced product for its time, using printed circuit boards, and what were then considered micro-miniature components.

Masaru Ibuka, co-founder of the Japanese firm Sony, was visiting the USA when Bell Labs announced the availability of manufacturing licenses, including detailed instructions on how to manufacture junction transistors. Ibuka obtained special permission from the Japanese Ministry of Finance to pay the $50,000 license fee, and in 1955 the company introduced their own five-transistor "coatpocket" radio, the TR-55, under the new brand name Sony. This product was soon followed by more ambitious designs, but it is generally regarded as marking the commencement of Sony's growth into a manufacturing superpower.

The TR-55 was quite similar to the Regency TR-1 in many ways, being powered by the same sort of 22.5-volt battery, and was not much more practical. Note: according to the schematic, the TR-55 used a 6 volt supply. Very few were distributed outside Japan. It was not until 1957 that Sony produced their ground-breaking "TR-63" shirt pocket radio, a much more advanced design that ran on a standard 9-volt battery and could compete favorably with vacuum tube portables. The TR-63 was also the first transistor radio to use all miniature components. (The term "pocket" was a matter of some interpretation, as Sony allegedly had special shirts made with over-sized pockets for their salesmen.)

Chrysler Mopar 914HR – First Transistor Car Radio

In the April 28th 1955 edition of the Wall Street Journal, Chrysler and Philco announced that they had developed and produced the world's first all-transistor car radio. Chrysler made the all-transistor car radio, Mopar model 914HR, available as an "option" in Fall 1955 for its new line of 1956 Chrysler and Imperial cars, which hit the showroom floor on October 21, 1955. The all-transistor car radio was a $150 option.

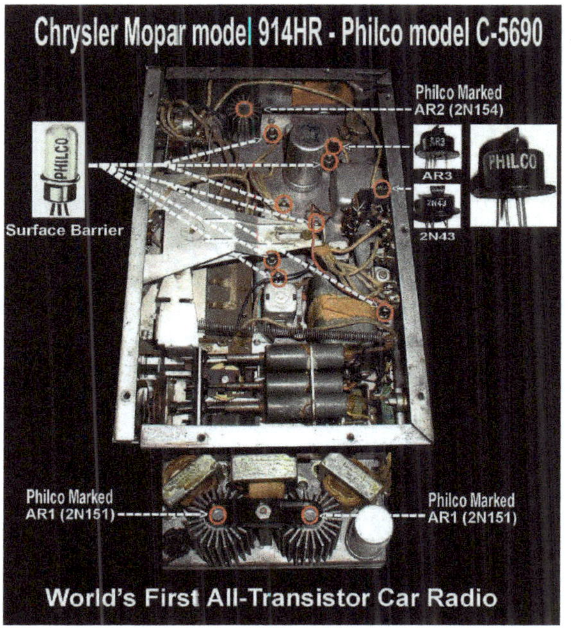

All-transistor car radio – Chrysler Mopar model 914HR

Hobby Use

The first low-cost junction transistor available to the general public was the CK722, a PNP germanium small signal unit introduced by Raytheon in early 1953 for $7.60 each. In the 1950s and 1960s, hundreds of hobbyist electronics projects based around the CK722 transistor were published in popular books and magazines. Raytheon also participated in expanding the role of the CK722 as a hobbyist electronics device by publishing "Transistor Applications" and "Transistor Applications- Volume 2" during the mid-1950s.

Computers

The world's first transistor computer was built at the University of Manchester in November 1953. The computer was built by Dick Grimsdale, then a research student in the Department of Electrical Engineering and later a Professor of Electronic Engineering at Sussex University. The machine used point-contact transistors, made in small quantities by STC and Mullard. These consisted of a single crystal of germanium with two fine wires, resembling the crystal and cat's whisker of the 1920s. These transistors had the useful property that a single transistor could possess two stable states. ... The development of the machine was severely hampered by the unreliability of the transistors, which consumed 150 watts.

Metropolitan Vickers Ltd rebuilt the full 200 transistor (& 1300 diode) design in 1956 using junction transistors (for internal use).

The IBM 7070 (1958), IBM 7090 (1959), and CDC 1604 (1960) were the first computers (as products for sale) based on transistors.

Improvements in Transistor Design

Shockley was upset about the device being credited to Brattain and Bardeen, who he felt had built it "behind his back" to take the glory. Matters became worse when Bell Labs lawyers found that some of Shockley's own writings on the transistor were close enough to those of an earlier 1925 patent by Julius Edgar Lilienfeld that they thought it best that his name be left off the patent application.

Shockley was incensed, and decided to demonstrate who was the real brains of the operation. Only a few months later he invented an entirely new type of transistor with a layer or "sandwich" structure. This new form was considerably more robust than the fragile point-contact system, and would go on to be used for the vast majority of all transistors into the 1960s. It would evolve into the bipolar junction transistor.

Further developments included the grown-junction transistor (1951), the surface-barrier transistor (1953), the diffusion transistor, the tetrode transistor, and the pentode transistor. The diffused silicon 'mesa transistor' was developed at Bell in 1955 and made commercially available by Fairchild Semiconductor in 1958. The spacistor was a type of transistor developed in the 1950s as an improvement over the point-contact transistor and the later alloy junction transistor.

Philco surface-barrier transistor developed and produced in 1953

First high frequency transistor developed by Philco in 1953

In 1953, Philco had developed the world's first high frequency surface-barrier transistor, which was also the first transistor that was suitable for high speed computers. The world's first all-transistor car radio, manufactured by Philco in 1955, used surface-barrier transistors in its circuitry design.

With the fragility problems solved, a remaining problem was purity. Making germanium of the required purity was proving to be a serious problem, and limited the number of transistors that actually worked from a given batch of material. Germanium's sensitivity to temperature also limited its usefulness.

Silicon

Scientists theorized that silicon would be easier to fabricate, but few bothered to investigate this possibility. Morris Tanenbaum et al. at Bell Laboratories were the first to develop a working silicon transistor on January 26, 1954. A few months later, Gordon Teal, working independently at the nascent Texas Instruments, developed a similar device. Both of these devices were made by controlling the doping of single silicon crystals while they were grown from molten silicon. A far superior method was developed by Morris Tanenbaum and Calvin S. Fuller at Bell Laboratories in early 1955 by the gaseous diffusion of donor and acceptor impurities into single crystal silicon chips.

The first gallium-arsenide Schottky-gate field-effect transistor (MESFET) was made by Carver Mead and reported in 1966.

IMPORTANCE

The transistor is the key active component in practically all modern electronics. Many consider it to be one of the greatest inventions of the 20th century. Its importance in today's society rests on its ability to be mass-produced using a highly automated process (semiconductor device fabrication) that achieves astonishingly low per-transistor costs. The invention of the first transistor at Bell Labs was named an IEEE Milestone in 2009.

A Darlington transistor opened up so the actual transistor chip (the small square) can be seen inside. A Darlington transistor is effectively two transistors on the same chip. One transistor is much larger than the other, but both are large in comparison to transistors in large-scale integration because this particular example is intended for power applications.

Although several companies each produce over a billion individually packaged (known as *discrete*) transistors every year, the vast majority of transistors are now produced in integrated circuits (often shortened to *IC*, *microchips* or simply *chips*), along with diodes, resistors, capacitors and other electronic components, to produce complete electronic circuits. A logic gate consists of up to about twenty transistors whereas an advanced microprocessor, as of 2009, can use as many as 3 billion transistors (MOSFETs). "About 60 million transistors were built in 2002... for [each] man, woman, and child on Earth."

The transistor's low cost, flexibility, and reliability have made it a ubiquitous device. Transistorized mechatronic circuits have replaced electromechanical devices in controlling appliances and machinery. It is often easier and cheaper to use a standard microcontroller and write a computer program to carry out a control function than to design an equivalent mechanical system to control that same function.

SIMPLIFIED OPERATION

The essential usefulness of a transistor comes from its ability to use a small signal applied between one pair of its terminals to control a much larger signal at another pair of terminals. This property is called gain. It can produce a stronger output signal, a voltage or current, which is proportional to a weaker input signal; that is, it can act as an amplifier. Alternatively, the transistor can be used to turn current on or off in a circuit as an electrically controlled switch, where the amount of current is determined by other circuit elements.

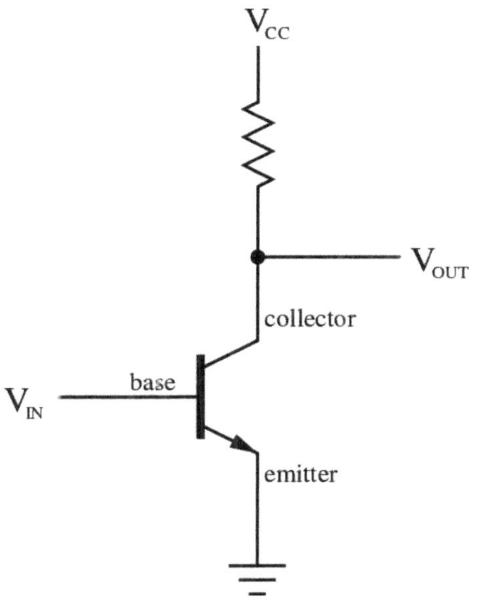

A simple circuit diagram to show the labels of a n–p–n bipolar transistor.

There are two types of transistors, which have slight differences in how they are used in a circuit. A *bipolar transistor* has terminals labeled base, collector, and emitter. A small current at the base terminal (that is, flowing between the base and the emitter) can control or switch a much larger current between the collector and emitter terminals. For a *field-effect transistor*, the terminals are labeled gate, source, and drain, and a voltage at the gate can control a current between source and drain.

The image represents a typical bipolar transistor in a circuit. Charge will flow between emitter and collector terminals depending on the current in the base. Because internally the base and emitter connections behave like a semiconductor diode, a voltage drop develops between base and emitter while the base current exists. The amount of this voltage depends on the material the transistor is made from, and is referred to as V_{BE}.

Transistor as a Switch

Transistors are commonly used in digital circuits as electronic switches which can be either in an "on" or "off" state, both for high-power applications such as switched-mode power supplies and for low-power applications such as logic gates. Important parameters for this application include the current switched, the voltage handled, and the switching speed, characterized by the rise and fall times.

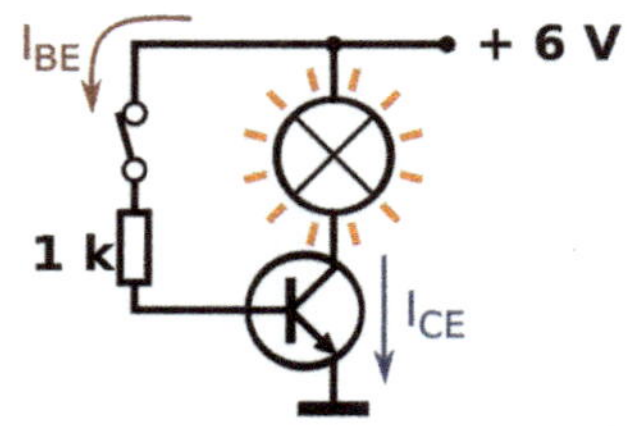

BJT used as an electronic switch, in grounded-emitter configuration.

In a grounded-emitter transistor circuit, such as the light-switch circuit shown, as the base voltage rises, the emitter and collector currents rise exponentially. The collector voltage drops because of reduced resistance from collector to emitter. If the voltage difference between the collector and emitter were zero (or near zero), the collector current would be limited only by the load resistance (light bulb) and the supply voltage. This is called *saturation* because current is flowing from collector to emitter freely. When saturated, the switch is said to be *on*.

Providing sufficient base drive current is a key problem in the use of bipolar transistors as switches. The transistor provides current gain, allowing a relatively large current in the collector to be switched by a much smaller current into the base terminal. The ratio of these currents varies depending on the type of transistor, and even for a particular type, varies depending on the collector current. In the example light-switch circuit shown, the resistor is chosen to provide enough base current to ensure the transistor will be saturated.

In a switching circuit, the idea is to simulate, as near as possible, the ideal switch having the properties of open circuit when off, short circuit when on, and an instantaneous transition between the two states. Parameters are chosen such that the "off" output is limited to leakage currents too small to affect connected circuitry; the resistance of the transistor in the "on" state is too small to affect circuitry; and the transition between the two states is fast enough not to have a detrimental effect.

Transistor as An Amplifier

The common-emitter amplifier is designed so that a small change in voltage (V_{in}) changes the small current through the base of the transistor; the transistor's current amplification combined with the properties of the circuit mean that small swings in V_{in} produce large changes in V_{out}.

Various configurations of single transistor amplifier are possible, with some providing current gain, some voltage gain, and some both.

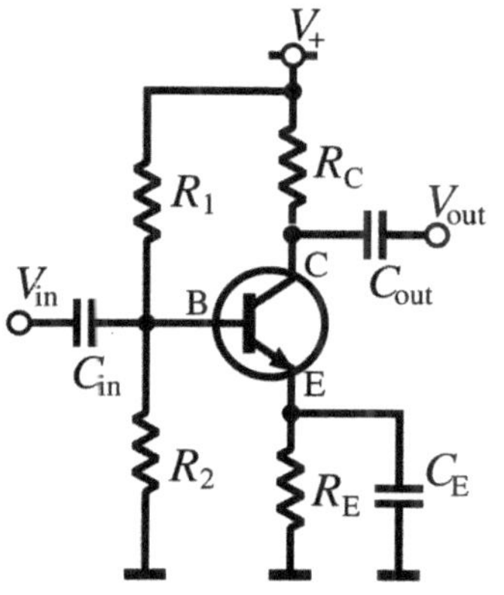

Amplifier circuit, common-emitter configuration with a voltage-divider bias circuit.

From mobile phones to televisions, vast numbers of products include amplifiers for sound reproduction, radio transmission, and signal processing. The first discrete-transistor audio amplifiers barely supplied a few hundred milliwatts, but power and audio fidelity gradually increased as better transistors became available and amplifier architecture evolved.

Modern transistor audio amplifiers of up to a few hundred watts are common and relatively inexpensive.

COMPARISON WITH VACUUM TUBES

Before transistors were developed, vacuum (electron) tubes (or in the UK "thermionic valves" or just "valves") were the main active components in electronic equipment.

Advantages

The key advantages that have allowed transistors to replace vacuum tubes in most applications are

- no cathode heater (which produces the characteristic orange glow of tubes), reducing power consumption, eliminating delay as tube heaters warm up, and immune from cathode poisoning and depletion;
- very small size and weight, reducing equipment size;
- large numbers of extremely small transistors can be manufactured as a single integrated circuit;
- low operating voltages compatible with batteries of only a few cells;
- circuits with greater energy efficiency are usually possible. For low-power applications (e.g., voltage amplification) in particular, energy consumption can be very much less than for tubes;
- inherent reliability and very long life; tubes always degrade and fail over time. Some transistorized devices have been in service for more than 50 years ;

- complementary devices available, providing design flexibility including complementary-symmetry circuits, not possible with vacuum tubes;
- very low sensitivity to mechanical shock and vibration, providing physical ruggedness and virtually eliminating shock-induced spurious signals (e.g., microphonics in audio applications);
- not susceptible to breakage of a glass envelope, leakage, outgassing, and other physical damage.

Limitations

Transistors have the following limitations:

- silicon transistors can age and fail;
- high-power, high-frequency operation, such as that used in over-the-air television broadcasting, is better achieved in vacuum tubes due to improved electron mobility in a vacuum;
- solid-state devices are susceptible to damage from very brief electrical and thermal events, including electrostatic discharge in handling; vacuum tubes are electrically much more rugged;
- sensitivity to radiation and cosmic rays (special radiation-hardened chips are used for spacecraft devices);
- vacuum tubes in audio applications create significant lower-harmonic distortion, the so-called tube sound, which some people prefer.

TYPES

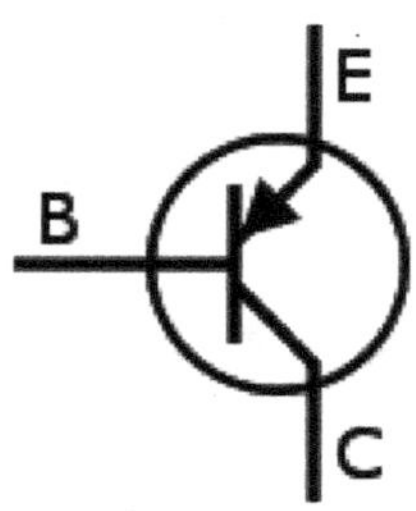

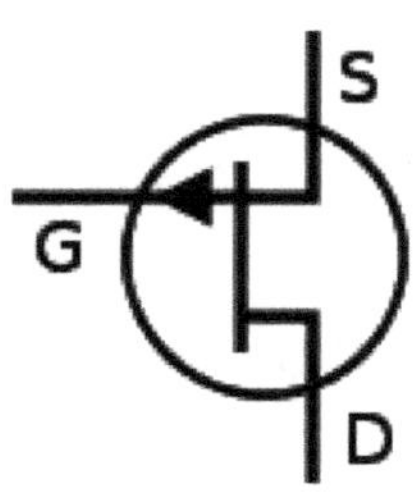

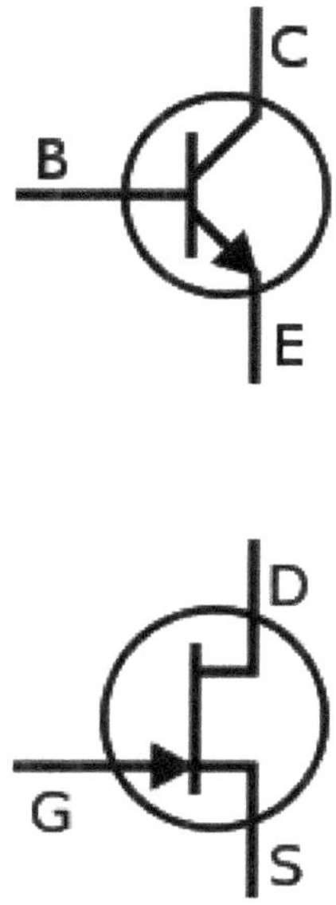

BJT and JFET symbols

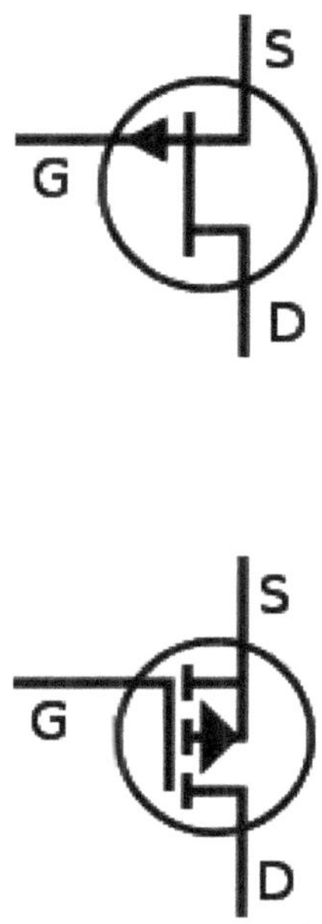

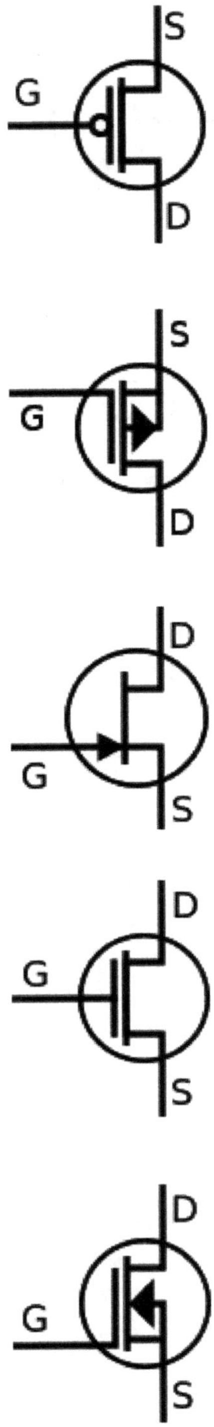

JFET and MOSFET symbols

Transistors are categorized by

- semiconductor material: the metalloids germanium (first used in 1947) and silicon (first used in 1954)—in amorphous, polycrystalline and monocrystalline form—, the compounds gallium arsenide (1966) and silicon carbide (1997), the alloy silicon-germanium (1989), the allotrope of carbon graphene (research ongoing since 2004), etc.;
- structure: BJT, JFET, IGFET (MOSFET), insulated-gate bipolar transistor, "other types";
- electrical polarity (positive and negative): n–p–n, p–n–p (BJTs), n-channel, p-channel (FETs);
- maximum power rating: low, medium, high;
- maximum operating frequency: low, medium, high, radio (RF), microwave frequency (the maximum effective frequency of a transistor in a common-emitter or common-source circuit is denoted by the term f_T, an abbreviation for transition frequency—the frequency of transition is the frequency at which the transistor yields unity voltage gain)
- application: switch, general purpose, audio, high voltage, super-beta, matched pair;
- physical packaging: through-hole metal, through-hole plastic, surface mount, ball grid array, power modules;
- amplification factor h_{FE}, β_F (transistor beta) or g_m (transconductance).

Hence, a particular transistor may be described as *silicon, surface-mount, BJT, n–p–n, low-power, high-frequency switch*.

A popular way to remember which symbol represents which type of transistor is to look at the arrow and how it is arranged. Within an NPN transistor symbol, the arrow will Not Point iN. Conversely, within the PNP symbol you see that the arrow Points iN Proudly.

Bipolar Junction Transistor (BJT)

A bipolar junction transistor (bipolar transistor or BJT) is a type of transistor that uses both electron and hole charge carriers. In contrast, unipolar transistors, such as field-effect transistors, only use one kind of charge carrier. For their operation, BJTs use two junctions between two semiconductor types, n-type and p-type.

BJTs are manufactured in two types, NPN and PNP, and are available as individual components, or fabricated in integrated circuits, often in large numbers. The basic function of a BJT is to amplify current. This allows BJTs to be used as amplifiers or switches, giving them wide applicability in electronic equipment, including computers, televisions, mobile phones, audio amplifiers, industrial control, and radio transmitters.

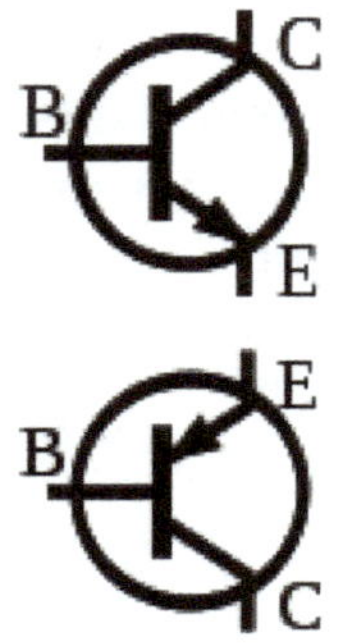

BJT schematic symbols

Note on Current Direction

By convention, the direction of current on diagrams is shown as the direction that a positive charge would move. This is called *conventional current*. However, current in metal conductors is due to the flow of electrons which, because they carry a negative charge, move in the opposite direction to conventional current. On the other hand, inside a bipolar transistor, currents can be composed of both positively charged holes and negatively charged electrons. In this article, current arrows are shown in the conventional direction, but labels for the movement of holes and electrons show their actual direction inside the transistor.

Function

BJTs come in two types, or polarities, known as PNP and NPN based on the doping types of the three main terminal regions. An NPN transistor comprises two semiconductor junctions that share a thin p-doped anode region, and a PNP transistor comprises two semiconductor junctions that share a thin n-doped cathode region.

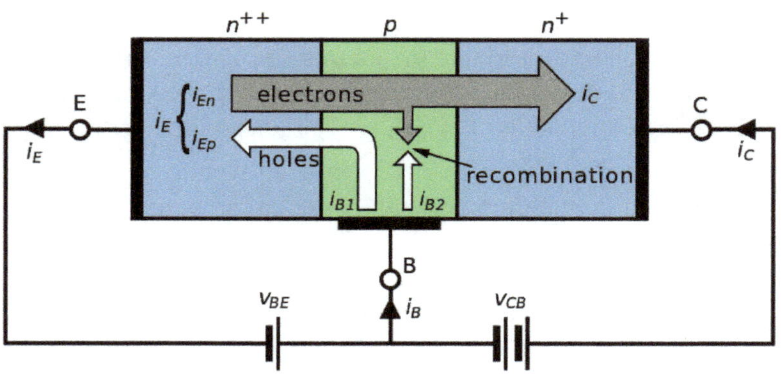

NPN BJT with forward-biased E–B junction and reverse-biased B–C junction

Charge flow in a BJT is due to diffusion of charge carriers across a junction between two regions of different charge concentrations. The regions of a BJT are called *emitter*, *collector*, and *base*. A discrete transistor has three leads for connection to these regions. Typically, the emitter region is heavily doped compared to the other two layers, whereas the majority charge carrier concentrations in base and collector layers are about the same. By design, most of the BJT collector current is due to the flow of charges injected from a high-concentration emitter into the base where they are minority carriers that diffuse toward the collector, and so BJTs are classified as minority-carrier devices.

In typical operation, the base–emitter junction is forward biased, which means that the p-doped side of the junction is at a more positive potential than the n-doped side, and the base–collector junction is reverse biased. In an NPN transistor, when positive bias is applied to the base–emitter junction, the equilibrium is disturbed between the thermally generated carriers and the repelling electric field of the n-doped emitter depletion region. This allows thermally excited electrons to inject from the emitter into the base region. These electrons diffuse through the base from the region of high concentration near the emitter towards the region of low concentration near the collector. The electrons in the base are called *minority carriers* because the base is doped p-type, which makes holes the *majority carrier* in the base.

To minimize the percentage of carriers that recombine before reaching the collector–base junction, the transistor's base region must be thin enough that carriers can diffuse across it in much less time than the semiconductor's minority carrier lifetime. In particular, the thickness of the base must be much less than the diffusion length of the electrons. The collector–base junction is reverse-biased, and so little electron injection occurs from the collector to the base, but electrons that diffuse through the base towards the collector are swept into the collector by the electric field in the depletion region of the collector–base junction. The thin *shared* base and asymmetric collector–emitter doping are what differentiates a bipolar transistor from two *separate* and oppositely biased diodes connected in series.

Voltage, Current, and Charge Control

The collector–emitter current can be viewed as being controlled by the base–emitter current (current control), or by the base–emitter voltage (voltage control). These views are related by the current–voltage relation of the base–emitter junction, which is just the usual exponential current–voltage curve of a p-n junction (diode).

The physical explanation for collector current is the concentration of minority carriers in the base region. Due to low level injection (in which there are much fewer excess carriers than normal majority carriers) the ambipolar transport rates (in which the excess majority and minority carriers flow at the same rate) is in effect determined by the excess minority carriers.

Detailed transistor models of transistor action, such as the Gummel–Poon model, account for the distribution of this charge explicitly to explain transistor behaviour more exactly. The charge-control view easily handles phototransistors, where minority carriers in the base region are created by the absorption of photons, and handles the dynamics of turn-off, or recovery time, which depends on charge in the base region recombining.

However, because base charge is not a signal that is visible at the terminals, the current- and voltage-control views are generally used in circuit design and analysis.

In analog circuit design, the current-control view is sometimes used because it is approximately linear. That is, the collector current is approximately β_F times the base current. Some basic circuits can be designed by assuming that the emitter–base voltage is approximately constant, and that collector current is beta times the base current. However, to accurately and reliably design production BJT circuits, the voltage-control (for example, Ebers–Moll) model is required. The voltage-control model requires an exponential function to be taken into account, but when it is linearized such that the transistor can be modeled as a transconductance, as in the Ebers–Moll model, design for circuits such as differential amplifiers again becomes a mostly linear problem, so the voltage-control view is often preferred. For translinear circuits, in which the exponential I–V curve is key to the operation, the transistors are usually modeled as voltage-controlled current sources whose transconductance is proportional to their collector current. In general, transistor-level circuit design is performed using SPICE or a comparable analog circuit simulator, so model complexity is usually not of much concern to the designer.

Turn-On, Turn-Off, and Storage Delay

The bipolar transistor exhibits a few delay characteristics when turning on and off. Most transistors, and especially power transistors, exhibit long base-storage times that limit maximum frequency of operation in switching applications. One method for reducing this storage time is by using a Baker clamp.

Transistor Parameters: Alpha (α) and Beta (β)

The proportion of electrons able to cross the base and reach the collector is a measure of the BJT efficiency. The heavy doping of the emitter region and light doping of the base region causes many more electrons to be injected from the emitter into the base than holes to be injected from the base into the emitter.

The *common-emitter current gain* is represented by β_F or the h-parameter h_{FE}; it is approximately the ratio of the DC collector current to the DC base current in forward-active region. It is typically greater than 50 for small-signal transistors but can be smaller in transistors designed for high-power applications.

Another important parameter is the *common-base current gain*, α_F. The common-base current gain is approximately the gain of current from emitter to collector in the forward-active region. This ratio usually has a value close to unity; between 0.980 and 0.998. It is less than unity due to recombination of charge carriers as they cross the base region.

Alpha and beta are more precisely related by the following identities (NPN transistor):

$$\alpha_F = \frac{I_{\mathrm{C}}}{I_{\mathrm{E}}}, \qquad \beta_F = \frac{I_{\mathrm{C}}}{I_{\mathrm{B}}}$$

$$\alpha_F = \frac{\beta_F}{1+\beta_F} \iff \beta_F = \frac{\alpha_F}{1-\alpha_F}$$

Structure

A BJT consists of three differently doped semiconductor regions: the *emitter* region, the *base* region and the *collector* region. These regions are, respectively, *p* type, *n* type and *p* type in a PNP transistor, and *n* type, *p* type and *n* type in an NPN transistor. Each semiconductor region is connected to a terminal, appropriately labeled: *emitter* (E), *base* (B) and *collector* (C).

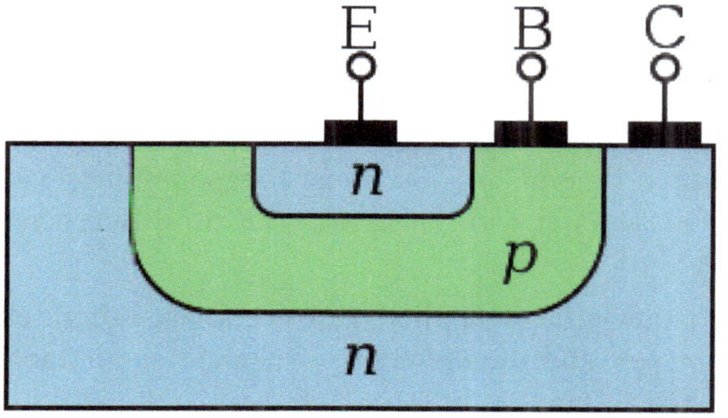

Simplified cross section of a planar *NPN* bipolar junction transistor

The *base* is physically located between the *emitter* and the *collector* and is made from lightly doped, high-resistivity material. The collector surrounds the emitter region, making it almost impossible for the electrons injected into the base region to escape without being collected, thus making the resulting value of α very close to unity, and so, giving the transistor a large β. A cross-section view of a BJT indicates that the collector–base junction has a much larger area than the emitter–base junction.

The bipolar junction transistor, unlike other transistors, is usually not a symmetrical device. This means that interchanging the collector and the emitter makes the transistor leave the forward active mode and start to operate in reverse mode. Because the transistor›s internal structure is usually optimized for forward-mode operation, interchanging the collector and the emitter makes the values of α and β in reverse operation much smaller than those in forward operation; often the α of the reverse mode is lower than 0.5. The lack of symmetry is primarily due to the doping ratios of the emitter and the collector. The emitter is heavily doped, while the collector is lightly doped, allowing a large reverse bias voltage to be applied before the collector–base junction breaks down. The collector–base junction is reverse biased in normal operation. The reason the emitter is heavily doped is to increase the emitter injection efficiency: the ratio of carriers injected by the emitter to those injected by the base. For high current gain, most of the carriers injected into the emitter–base junction must come from the emitter.

Die of a KSY34 high-frequency NPN transistor. Bond wires connect to the base and emitter

The low-performance "lateral" bipolar transistors sometimes used in CMOS processes are sometimes designed symmetrically, that is, with no difference between forward and backward operation.

Small changes in the voltage applied across the base–emitter terminals cause the current between the *emitter* and the *collector* to change significantly. This effect can be used to amplify the input voltage or current. BJTs can be thought of as voltage-controlled current sources, but are more simply characterized as current-controlled current sources, or current amplifiers, due to the low impedance at the base.

Early transistors were made from germanium but most modern BJTs are made from silicon. A significant minority are also now made from gallium arsenide, especially for very high speed applications.

NPN

NPN is one of the two types of bipolar transistors, consisting of a layer of P-doped semiconductor (the "base") between two N-doped layers. A small current entering the base is amplified to produce a large collector and emitter current. That is, when there is a positive potential difference measured from the base of an NPN transistor to its emitter (that is, when the base is **high** relative to the emitter), as well as a positive potential difference measured from the collector to the emitter, the transistor becomes active. In this "on" state, charge flows from the collector to the emitter of the transistor. Most of the current is carried by electrons moving from emitter to collector as minority carriers in the P-type base region. To allow for greater current and faster operation, most bipolar transistors used today are NPN because electron mobility is higher than hole mobility.

The symbol of an NPN BJT. A mnemonic for the symbol is "**n**ot **p**ointing i**n**".

A mnemonic device for the NPN transistor symbol is "*n*ot *p*ointing i*n*", based on the arrows in the symbol and the letters in the name.

PNP

The other type of BJT is the PNP, consisting of a layer of N-doped semiconductor between two layers of P-doped material. A small current leaving the base is amplified in the collector output. That is, a PNP transistor is "on" when its base is pulled low relative to the emitter. In a PNP transistor, the emitter–base region is forward biased, so holes are injected into the base as minority carriers. The base is very thin, and most of the holes cross the reverse-biased base–collector junction to the collector.

The symbol of a PNP BJT. A mnemonic for the symbol is "points in proudly".

The arrows in the NPN and PNP transistor symbols are on the emitter legs and point in the direction of the conventional current when the device is in forward active or forward saturated mode.

A mnemonic device for the PNP transistor symbol is "*p*ointing i*n* (*p*roudly/*p*ermanently)", based on the arrows in the symbol and the letters in the name.

Heterojunction Bipolar Transistor

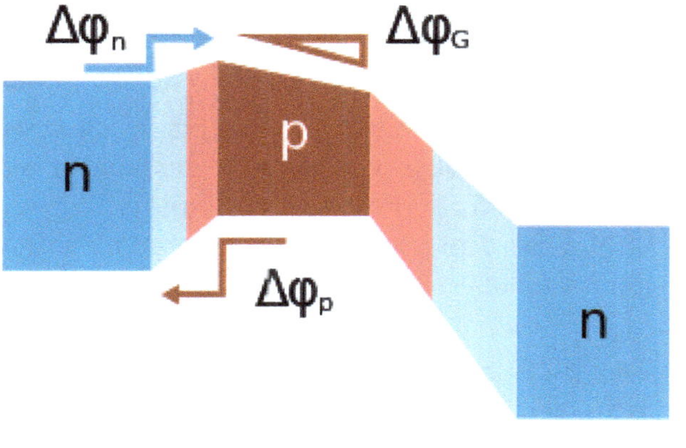

Bands in graded heterojunction NPN bipolar transistor. Barriers indicated for electrons to move from emitter to base, and for holes to be injected backward from base to emitter; also, grading of bandgap in base assists electron transport in base region; light colors indicate depleted regions.

The heterojunction bipolar transistor (HBT) is an improvement of the BJT that can handle signals of very high frequencies up to several hundred GHz. It is common in modern ultrafast circuits, mostly RF systems.

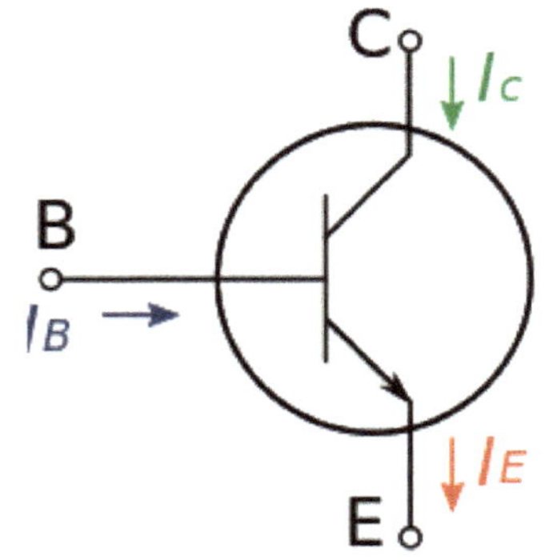

Symbol for NPN Bipolar Transistor with current flow direction.

Heterojunction transistors have different semiconductors for the elements of the transistor. Usually the emitter is composed of a larger bandgap material than the base. The figure shows that this difference in bandgap allows the barrier for holes to inject backward from the base into the emitter, denoted in the figure as $\Delta\varphi_p$, to be made large, while the barrier for electrons to inject into the base $\Delta\varphi_n$ is made low. This barrier arrangement helps reduce minority carrier injection from the base when the emitter-base junction is under forward bias, and thus reduces base current and increases emitter injection efficiency.

The improved injection of carriers into the base allows the base to have a higher doping level, resulting in lower resistance to access the base electrode. In the more traditional BJT, also referred to as homojunction BJT, the efficiency of carrier injection from the emitter to the base is primarily determined by the doping ratio between the emitter and base, which means the base must be lightly doped to obtain high injection efficiency, making its resistance relatively high. In addition, higher doping in the base can improve figures of merit like the Early voltage by lessening base narrowing.

The grading of composition in the base, for example, by progressively increasing the amount of germanium in a SiGe transistor, causes a gradient in bandgap in the neutral base, denoted in the figure by $\Delta\varphi_G$, providing a "built-in" field that assists electron transport across the base. That drift component of transport aids the normal diffusive transport, increasing the frequency response of the transistor by shortening the transit time across the base.

Two commonly used HBTs are silicon–germanium and aluminum gallium arsenide, though a wide variety of semiconductors may be used for the HBT structure. HBT structures are usually grown by epitaxy techniques like MOCVD and MBE.

Field-Effect Transistor (FET)

The field-effect transistor (FET) is a transistor that uses an electric field to control the shape and hence the electrical conductivity of a channel of one type of charge carrier in a semiconductor material. FETs are also known as unipolar transistors since they involve single-carrier-type operation. The FET has several forms, but all have high input impedance. While the conductivity of a non-FET is regulated by the input current (the emitter to base current) and so has a low input impedance, a FET's conductivity is regulated by a voltage applied to a terminal (the gate) which is insulated from the device. The applied gate voltage imposes an electric field into the device, which in turn attracts or repels charge carriers to or from the region between a source terminal and a drain terminal. The density of charge carriers in turn influences the conductivity between the source and drain.

History

The field-effect transistor was first patented by Julius Edgar Lilienfeld in 1926 and by Oskar Heil in 1934, but practical semiconducting devices (the JFET) were developed only much later after the transistor effect was observed and explained by the team of William Shockley at Bell Labs in 1947, immediately after the 20-year patent period eventually expired. The MOSFET, which largely superseded the JFET and had a profound effect on digital electronic development, was invented by Dawon Kahng and Martin Atalla in 1959.

Basic Information

FETs can be majority-charge-carrier devices, in which the current is carried predominantly by majority carriers, or minority-charge-carrier devices, in which the current is mainly due to a flow of minority carriers. The device consists of an active channel through which charge carriers, electrons or holes, flow from the source to the drain. Source and drain terminal conductors are connected to the semiconductor through ohmic contacts. The conductivity of the channel is a function of the potential applied across the gate and source terminals.

The FET's three terminals are:

- Source (S), through which the carriers enter the channel. Conventionally, current entering the channel at S is designated by I_S.
- Drain (D), through which the carriers leave the channel. Conventionally, current entering the channel at D is designated by I_D. Drain-to-source voltage is V_{DS}.
- Gate (G), the terminal that modulates the channel conductivity. By applying voltage to G, one can control I_D.

More About Terminals

All FETs have *source*, *drain*, and *gate* terminals that correspond roughly to the *emitter*, *collector*, and *base* of BJTs. Most FETs have a fourth terminal called the *body*, *base*, *bulk*, or *substrate*. This fourth terminal serves to bias the transistor into operation; it is rare to make non-trivial use of the body terminal in circuit designs, but its presence is important when setting up the physical layout of an integrated circuit. The size of the gate, length *L* in the diagram, is the distance between source and drain. The *width* is the extension of the transistor, in the direction perpendicular to the cross section in the diagram (i.e., into/out of the screen). Typically the width is much larger than the length of the gate. A gate length of 1 μm limits the upper frequency to about 5 GHz, 0.2 μm to about 30 GHz.

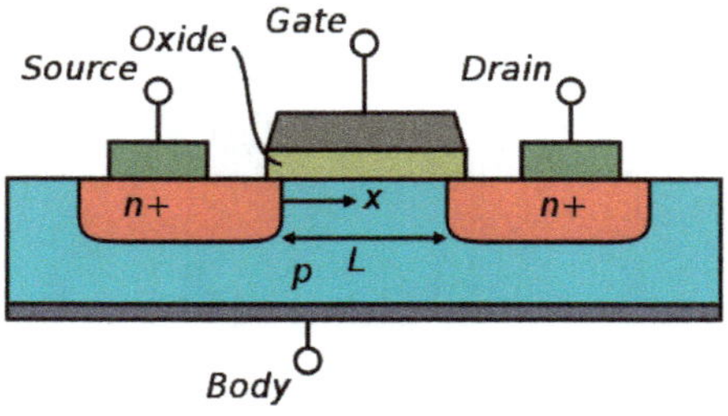

Cross section of an n-type MOSFET

The names of the terminals refer to their functions. The gate terminal may be thought of as controlling the opening and closing of a physical gate. This gate permits electrons to flow through or blocks their passage by creating or eliminating a channel between the source and drain. Electron-flow from the source terminal towards the drain terminal is influenced by an applied voltage. The body simply refers to the bulk of the semiconductor in which the gate, source and drain lie. Usually the body terminal is connected to the highest or lowest voltage within the circuit, depending on the type of the FET. The body terminal and the source terminal are sometimes connected together since the source is often connected to the highest or lowest voltage within the circuit, although there are several uses of FETs which do not have such a configuration, such as transmission gates and cascode circuits.

Effect of Gate Voltage On Current

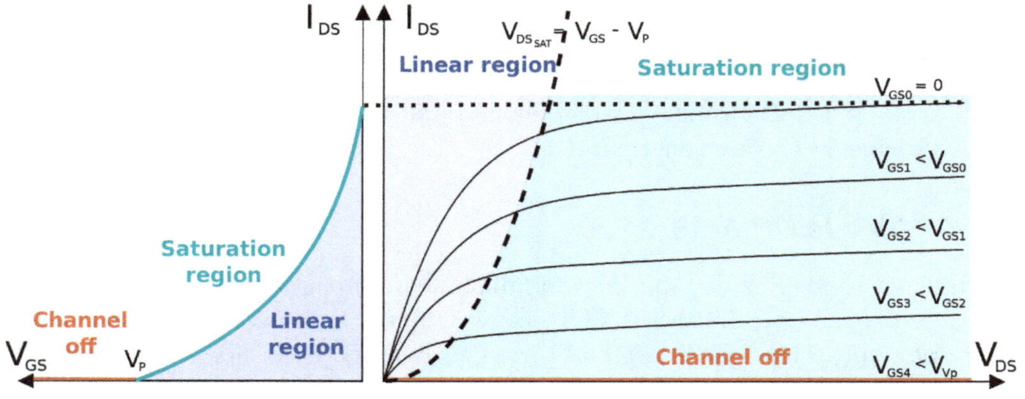

I–V characteristics and output plot of a JFET n-channel transistor.

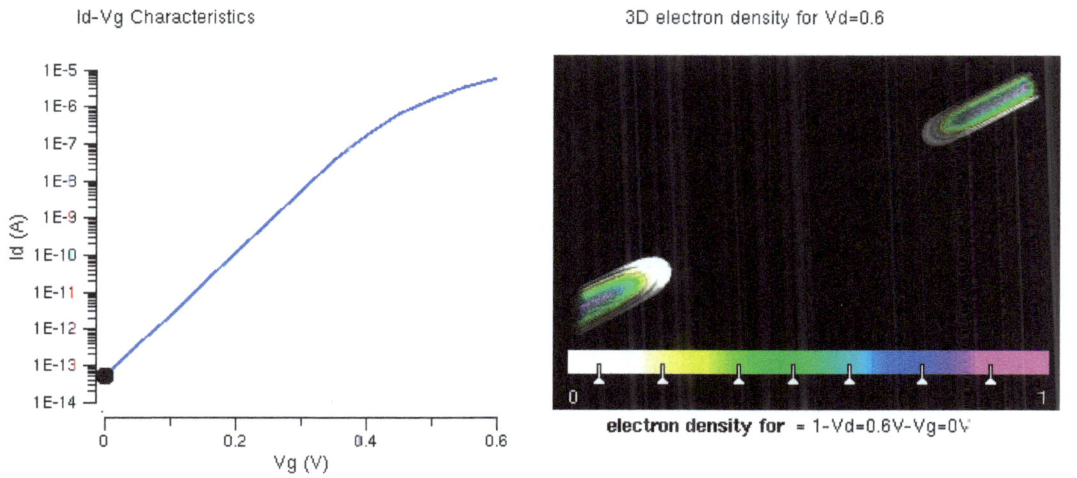

Simulation result for Right side: formation of inversion channel (electron density) and Left side: current-gate voltage curve(transfer characteristics) in a n-channel nanowire MOSFET. Note that the threshold voltage for this device lies around 0.45 V.

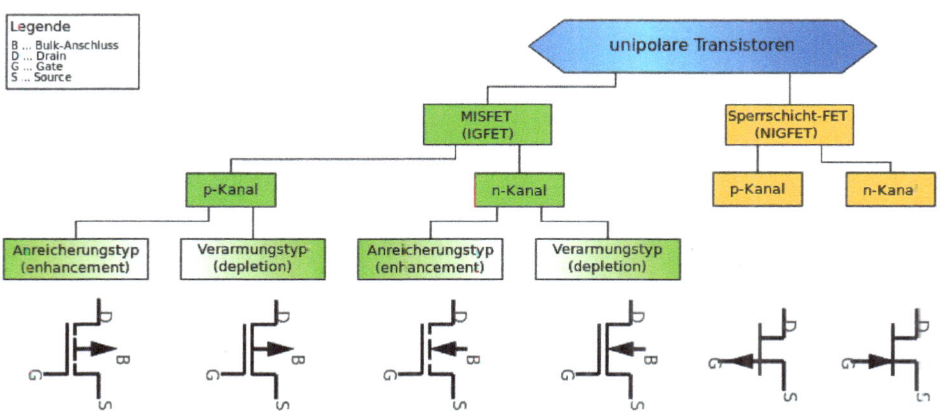

FET conventional symbol types

The FET controls the flow of electrons (or electron holes) from the source to drain by affecting the size and shape of a "conductive channel" created and influenced by voltage (or lack of voltage) applied across the gate and source terminals. (For simplicity, this discussion assumes that the body and source are connected.) This conductive channel is the "stream" through which electrons flow from source to drain.

n-Channel

In an n-channel *depletion-mode* device, a negative gate-to-source voltage causes a *depletion region* to expand in width and encroach on the channel from the sides, narrowing the channel. If the active region expands to completely close the channel, the resistance of the channel from source to drain becomes large, and the FET is effectively turned off like a switch. This is called *pinch-off,* and the voltage at which it occurs is called the *pinch-off voltage.* Conversely, a positive gate-to-source voltage increases the channel size and allows electrons to flow easily.

In an n-channel *enhancement-mode* device, a conductive channel does not exist naturally within the transistor, and a positive gate-to-source voltage is necessary to create one. The positive voltage attracts free-floating electrons within the body towards the gate, forming a conductive channel. But first, enough electrons must be attracted near the gate to counter the dopant ions added to the body of the FET; this forms a region with no mobile carriers called a depletion region, and the voltage at which this occurs is referred to as the *threshold voltage* of the FET. Further gate-to-source voltage increase will attract even more electrons towards the gate which are able to create a conductive channel from source to drain; this process is called *inversion.*

p-Channel

In a p-channel *depletion-mode* device, a positive voltage from gate to body creates a depletion layer by forcing the positively charged holes to the gate-insulator/semiconductor interface, leaving exposed a carrier-free region of immobile, negatively charged acceptor ions.

Effect of Source/Drain Voltage on Channel

For either enhancement- or depletion-mode devices, at drain-to-source voltages much less than gate-to-source voltages, changing the gate voltage will alter the channel resistance, and drain current will be proportional to drain voltage (referenced to source voltage). In this mode the FET operates like a variable resistor and the FET is said to be operating in a *linear mode* or *ohmic mode.*

If drain-to-source voltage is increased, this creates a significant asymmetrical change in the shape of the channel due to a gradient of voltage potential from source to drain. The shape of the inversion region becomes "pinched-off" near the drain end of the channel. If drain-to-source voltage is increased further, the pinch-off point of the channel begins to move away from the drain towards the source. The FET is said to be in *saturation mode*; although some authors refer to it as *active mode*, for a better analogy with bipolar transistor operating regions. The saturation mode, or the region between ohmic and saturation, is used when amplification is needed. The in-between region is sometimes considered to be part of the ohmic or linear region, even where drain current is not approximately linear with drain voltage.

Even though the conductive channel formed by gate-to-source voltage no longer connects source to drain during saturation mode, carriers are not blocked from flowing. Considering again an n-channel enhancement-mode device, a depletion region exists in the p-type body, surrounding the conductive channel and drain and source regions. The electrons which comprise the channel are free to move out of the channel through the depletion region if attracted to the drain by drain-to-source voltage. The depletion region is free of carriers and has a resistance similar to silicon. Any increase of the drain-to-source voltage will increase the distance from drain to the pinch-off point, increasing the resistance of the depletion region in proportion to the drain-to-source voltage applied. This proportional change causes the drain-to-source current to remain relatively fixed, independent of changes to the drain-to-source voltage, quite unlike its ohmic behavior in the linear mode of operation. Thus, in saturation mode, the FET behaves as a constant-current source rather than as a resistor, and can effectively be used as a voltage amplifier. In this case, the gate-to-source voltage determines the level of constant current through the channel.

Composition

FETs can be constructed from various semiconductors, with silicon being by far the most common. Most FETs are made by using conventional bulk semiconductor processing techniques, using a single crystal semiconductor wafer as the active region, or channel.

Among the more unusual body materials are amorphous silicon, polycrystalline silicon or other amorphous semiconductors in thin-film transistors or organic field-effect transistors (OFETs) that are based on organic semiconductors; often, OFET gate insulators and electrodes are made of organic materials, as well. Such FETs are manufactured using a variety of materials such as silicon carbide (SiC), gallium arsenide (GaAs), gallium nitride (GaN), and indium gallium arsenide (InGaAs). In June 2011, IBM announced that it had successfully used graphene-based FETs in an integrated circuit. These transistors are capable of about 2.23 GHz cutoff frequency, much higher than standard silicon FETs.

Types

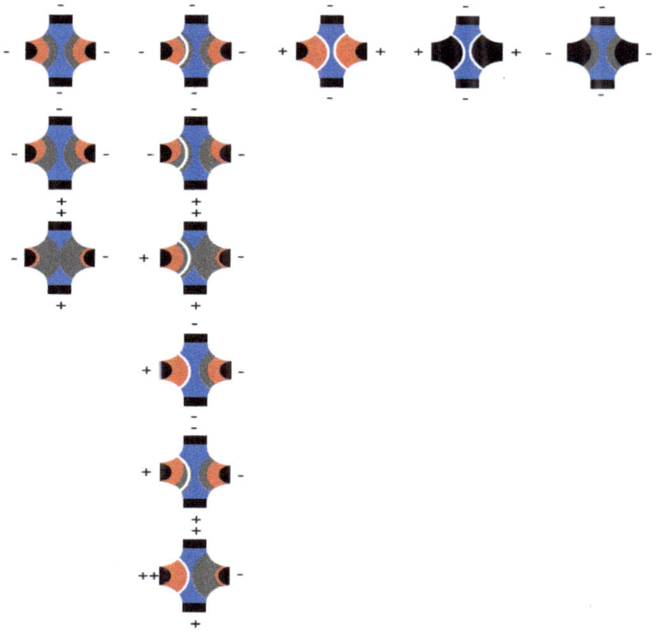

Depletion-type FETs under typical voltages: JFET, poly-silicon MOSFET, double-gate MOSFET, metal-gate MOSFET, MESFET.

- Depletion
- Electrons
- Holes
- Metal
- Insulator

Top: source, bottom: drain, left: gate, right: bulk. Voltages that lead to channel formation are not shown.

The channel of a FET is doped to produce either an n-type semiconductor or a p-type semiconductor. The drain and source may be doped of opposite type to the channel, in the case of enhancement mode FETs, or doped of similar type to the channel as in depletion mode FETs. Field-effect transistors are also distinguished by the method of insulation between channel and gate. Types of FETs include:

- The JFET (junction field-effect transistor) uses a reverse biased p–n junction to separate the gate from the body.
- The MOSFET (metal–oxide–semiconductor field-effect transistor) utilizes an insulator (typically SiO_2) between the gate and the body.
- The MNOS (metal–nitride–oxide–semiconductor) transistor utilizes an nitride-oxide layer insulator between the gate and the body.
- The DGMOSFET (dual-gate MOSFET) is a FET with two insulated gates.
- The DEPFET is a FET formed in a fully depleted substrate and acts as a sensor, amplifier and memory node at the same time. It can be used as an image (photon) sensor.
- The FREDFET (fast-reverse or fast-recovery epitaxial diode FET) is a specialized FET designed to provide a very fast recovery (turn-off) of the body diode.
- The HIGFET (heterostructure insulated gate field-effect transistor) is now used mainly in research.
- The MODFET (modulation-doped field-effect transistor) uses a quantum well structure formed by graded doping of the active region.
- The TFET (tunnel field-effect transistor) is based on band-to-band tunneling.
- The IGBT (insulated-gate bipolar transistor) is a device for power control. It has a structure akin to a MOSFET coupled with a bipolar-like main conduction channel. These are commonly used for the 200–3000 V drain-to-source voltage range of operation. Power MOSFETs are still the device of choice for drain-to-source voltages of 1 to 200 V.
- The HEMT (high-electron-mobility transistor), also called a HFET (heterostructure FET), can be made using bandgap engineering in a ternary semiconductor such as AlGaAs. The fully depleted wide-band-gap material forms the isolation between gate and body.

- The ISFET (ion-sensitive field-effect transistor) can be used to measure ion concentrations in a solution; when the ion concentration (such as H^+, see pH electrode) changes, the current through the transistor will change accordingly.
- The BioFET (Biologically sensitive field-effect transistor) is a class of sensors/biosensors based on ISFET technology which are utilized to detect charged molecules; when a charged molecule is present, changes in the electrostatic field at the BioFET surface result in a measurable change in current through the transistor. These include EnFETs, ImmunoFETs, GenFETs, DNAFETs, CPFETs, BeetleFETs, and FETs based on ion-channels/protein binding.
- The MESFET (metal–semiconductor field-effect transistor) substitutes the p–n junction of the JFET with a Schottky barrier; and is used in GaAs and other III-V semiconductor materials.
- The NOMFET is a nanoparticle organic memory field-effect transistor.
- The GNRFET (graphene nanoribbon field-effect transistor) uses a graphene nanoribbon for its channel.
- The VeSFET (vertical-slit field-effect transistor) is a square-shaped junctionless FET with a narrow slit connecting the source and drain at opposite corners. Two gates occupy the other corners, and control the current through the slit.
- The CNTFET (carbon nanotube field-effect transistor).
- The OFET (organic field-effect transistor) uses an organic semiconductor in its channel.
- The DNAFET (DNA field-effect transistor) is a specialized FET that acts as a biosensor, by using a gate made of single-strand DNA molecules to detect matching DNA strands.
- The QFET (quantum field effect transistor) takes advantage of quantum tunneling to greatly increase the speed of transistor operation by eliminating the traditional transistor's area of electron conduction.

Advantages

One advantage of the FET is its high gate to main current resistance, on the order of 100 MΩ or more, thus providing a high degree of isolation between control and flow. Because base current noise will increase with shaping time, a FET typically produces less noise than a bipolar junction transistor (BJT), and is thus found in noise sensitive electronics such as tuners and low-noise amplifiers for VHF and satellite receivers. It is relatively immune to radiation. It exhibits no offset voltage at zero drain current and hence makes an excellent signal chopper. It typically has better thermal stability than a BJT. Because they are controlled by gate charge, once the gate is closed or opened, there is no additional power draw, as there would be with a bipolar junction transistor or with non-latching relays in some states. This allows extremely low-power switching, which in turn allows greater miniaturization of circuits because heat dissipation needs are reduced compared to other types of switches.

Disadvantages

It has a relatively low gain–bandwidth product compared to a BJT. The MOSFET has a drawback of being very susceptible to overload voltages, thus requiring special handling during installation. The fragile insulating layer of the MOSFET between the gate and channel makes it vulnerable to electrostatic damage or changes to threshold voltage during handling. This is not usually a problem after the device has been installed in a properly designed circuit.

FETs often have a very low 'on' resistance and have a high 'off' resistance. However the intermediate resistances are significant, and so FETs can dissipate large amounts of power while switching. Thus efficiency can put a premium on switching quickly, but this can cause transients that can excite stray inductances and generate significant voltages that can couple to the gate and cause unintentional switching. FET circuits can therefore require very careful layout and can involve trades between switching speed and power dissipation. There is also a trade-off between voltage rating and 'on' resistance, so high voltage FETs have a relatively high 'on' resistance and hence conduction losses.

Uses

The most commonly used FET is the MOSFET. The CMOS (complementary metal oxide semiconductor) process technology is the basis for modern digital integrated circuits. This process technology uses an arrangement where the (usually "enhancement-mode") p-channel MOSFET and n-channel MOSFET are connected in series such that when one is ON, the other is OFF.

In FETs, electrons can flow in either direction through the channel when operated in the linear mode. The naming convention of drain terminal and source terminal is somewhat arbitrary, as the devices are typically (but not always) built symmetrically from source to drain. This makes FETs suitable for switching analog signals between paths (multiplexing). With this concept, one can construct a solid-state mixing board, for example.

A common use of the FET is as an amplifier. For example, due to its large input resistance and low output resistance, it is effective as a buffer in common-drain (source follower) configuration.

IGBTs are used in switching internal combustion engine ignition coils, where fast switching and voltage blocking capabilities are important.

Source-Gated Transistor

Source-gated transistors are more robust to manufacturing and environmental issues in large-area electronics such as display screens, but are slower in operation than FETs.

Usage of Bipolar and Field-Effect Transistors

The bipolar junction transistor (BJT) was the most commonly used transistor in the 1960s and 70s. Even after MOSFETs became widely available, the BJT remained the transistor of choice for many analog circuits such as amplifiers because of their greater linearity and ease of manufacture. In integrated circuits, the desirable properties of MOSFETs allowed them to capture nearly all market share for digital circuits. Discrete MOSFETs can be applied in transistor applications, including analog circuits, voltage regulators, amplifiers, power transmitters and motor drivers.

Other Transistor Types

Transistor symbol created on Portuguese pavement in the University of Aveiro.

- Bipolar junction transistor (BJT):
  - heterojunction bipolar transistor, up to several hundred GHz, common in modern ultrafast and RF circuits;
  - Schottky transistor;
  - avalanche transistor:
  - Darlington transistors are two BJTs connected together to provide a high current gain equal to the product of the current gains of the two transistors;
  - insulated-gate bipolar transistors (IGBTs) use a medium-power IGFET, similarly connected to a power BJT, to give a high input impedance. Power diodes are often connected between certain terminals depending on specific use. IGBTs are particularly suitable for heavy-duty industrial applications. The ASEA Brown Boveri (ABB) *5SNA2400E170100* illustrates just how far power semiconductor technology has advanced. Intended for three-phase power supplies, this device houses three n–p–n IGBTs in a case measuring 38 by 140 by 190 mm and weighing 1.5 kg. Each IGBT is rated at 1,700 volts and can handle 2,400 amperes;
  - phototransistor;

- multiple-emitter transistor, used in transistor–transistor logic and integrated current mirrors;
- multiple-base transistor, used to amplify very-low-level signals in noisy environments such as the pickup of a record player or radio front ends. Effectively, it is a very large number of transistors in parallel where, at the output, the signal is added constructively, but random noise is added only stochastically.

■ Field-effect transistor (FET):

- carbon nanotube field-effect transistor (CNFET), where the channel material is replaced by a carbon nanotube;
- junction gate field-effect transistor (JFET), where the gate is insulated by a reverse-biased p–n junction;
- metal–semiconductor field-effect transistor (MESFET), similar to JFET with a Schottky junction instead of a p–n junction;
- high-electron-mobility transistor (HEMT);
- metal–oxide–semiconductor field-effect transistor (MOSFET), where the gate is insulated by a shallow layer of insulator;
- inverted-T field-effect transistor (ITFET);
- fin field-effect transistor (FinFET), source/drain region shapes fins on the silicon surface;
- fast-reverse epitaxial diode field-effect transistor (FREDFET);
- thin-film transistor, in LCDs;
- organic field-effect transistor (OFET), in which the semiconductor is an organic compound;
- ballistic transistor;
- floating-gate transistor, for non-volatile storage;
- FETs used to sense environment;
- ion-sensitive field-effect transistor (IFSET), to measure ion concentrations in solution,
- electrolyte–oxide–semiconductor field-effect transistor (EOSFET), neurochip,
- deoxyribonucleic acid field-effect transistor (DNAFET).

■ Tunnel field-effect transistor, where it switches by modulating quantum tunnelling through a barrier.

■ Diffusion transistor, formed by diffusing dopants into semiconductor substrate; can be both BJT and FET.

■ Unijunction transistor, can be used as simple pulse generators. It comprise a main body of either P-type or N-type semiconductor with ohmic contacts at each end (terminals *Base1* and *Base2*). A junction with the opposite semiconductor type is formed at a point along the length of the body for the third terminal (*Emitter*).

- Single-electron transistors (SET), consist of a gate island between two tunneling junctions. The tunneling current is controlled by a voltage applied to the gate through a capacitor.
- Nanofluidic transistor, controls the movement of ions through sub-microscopic, water-filled channels.
- Multigate devices:
    - tetrode transistor;
    - pentode transistor;
    - trigate transistor (prototype by Intel);
    - dual-gate field-effect transistors have a single channel with two gates in cascode; a configuration optimized for *high-frequency amplifiers*, *mixers*, and oscillators.
- Junctionless nanowire transistor (JNT), uses a simple nanowire of silicon surrounded by an electrically isolated "wedding ring" that acts to gate the flow of electrons through the wire.
- Vacuum-channel transistor, when in 2012, NASA and the National Nanofab Center in South Korea were reported to have built a prototype vacuum-channel transistor in only 150 nanometers in size, can be manufactured cheaply using standard silicon semiconductor processing, can operate at high speeds even in hostile environments, and could consume just as much power as a standard transistor.
- Organic electrochemical transistor.

PART NUMBERING STANDARDS/SPECIFICATIONS

The types of some transistors can be parsed from the part number. There are three major semiconductor naming standards; in each the alphanumeric prefix provides clues to type of the device.

Japanese Industrial Standard (JIS)

JIS Transistor Prefix Table	
Prefix	**Type of transistor**
2SA	high-frequency p–n–p BJTs
2SB	audio-frequency p–n–p BJTs
2SC	high-frequency n–p–n BJTs
2SD	audio-frequency n–p–n BJTs
2SJ	P-channel FETs (both JFETs and MOSFETs)
2SK	N-channel FETs (both JFETs and MOSFETs)

The *JIS-C-7012* specification for transistor part numbers starts with "2S", e.g. 2SD965, but sometimes the "2S" prefix is not marked on the package – a 2SD965 might only be marked "D965"; a 2SC1815 might be listed by a supplier as simply "C1815". This series sometimes has suffixes (such as "R", "O", "BL", standing for "red", "orange", "blue", etc.) to denote variants, such as tighter h_{FE} (gain) groupings.

European Electronic Component Manufacturers Association (EECA)

The Pro Electron standard, the European Electronic Component Manufacturers Association part numbering scheme, begins with two letters: the first gives the semiconductor type (A for germanium, B for silicon, and C for materials like GaAs); the second letter denotes the intended use (A for diode, C for general-purpose transistor, etc.). A 3-digit sequence number (or one letter then 2 digits, for industrial types) follows. With early devices this indicated the case type. Suffixes may be used, with a letter (e.g. "C" often means high h_{FE}, such as in: BC549C) or other codes may follow to show gain (e.g. BC327-25) or voltage rating (e.g. BUK854-800A). The more common prefixes are:

Pro Electron / EECA Transistor Prefix Table				
Prefix class	**Type and usage**	**Example**	**Equivalent**	**Reference**
AC	Germanium small-signal AF transistor	AC126	NTE102A	Datasheet
AD	Germanium AF power transistor	AD133	NTE179	Datasheet
AF	Germanium small-signal RF transistor	AF117	NTE160	Datasheet
AL	Germanium RF power transistor	ALZ10	NTE100	Datasheet
AS	Germanium switching transistor	ASY28	NTE101	Datasheet
AU	Germanium power switching transistor	AU103	NTE127	Datasheet
BC	Silicon, small-signal transistor ("general purpose")	BC548	2N3904	Datasheet
BD	Silicon, power transistor	BD139	NTE375	Datasheet
BF	Silicon, RF (high frequency) BJT or FET	BF245	NTE133	Datasheet
BS	Silicon, switching transistor (BJT or MOSFET)	BS170	2N7000	Datasheet
BL	Silicon, high frequency, high power (for transmitters)	BLW60	NTE325	Datasheet
BU	Silicon, high voltage (for CRT horizontal deflection circuits)	BU2520A	NTE2354	Datasheet
CF	Gallium Arsenide small-signal Microwave transistor (MESFET)	CF739	—	Datasheet
CL	Gallium Arsenide Microwave power transistor (FET)	CLY10	—	Datasheet

Joint Electron Devices Engineering Council (JEDEC)

The JEDEC *EIA370* transistor device numbers usually start with "2N", indicating a three-terminal device (dual-gate field-effect transistors are four-terminal devices, so begin with 3N), then a 2, 3 or 4-digit sequential number with no significance as to device properties (although early devices with low numbers tend to be germanium). For example, 2N3055 is a silicon n–p–n power transistor, 2N1301 is a p–n–p germanium switching transistor. A letter suffix (such as "A") is sometimes used to indicate a newer variant, but rarely gain groupings.

Proprietary

Manufacturers of devices may have their own proprietary numbering system, for example CK722. Since devices are second-sourced, a manufacturer's prefix (like "MPF" in MPF102, which originally would denote a Motorola FET) now is an unreliable indicator of who made the device. Some proprietary naming schemes adopt parts of other naming schemes, for example a PN2222A is a (possibly Fairchild Semiconductor) 2N2222A in a plastic case (but a PN108 is a plastic version of a BC108, not a 2N108, while the PN100 is unrelated to other xx100 devices).

Military part numbers sometimes are assigned their own codes, such as the British Military CV Naming System.

Manufacturers buying large numbers of similar parts may have them supplied with "house numbers", identifying a particular purchasing specification and not necessarily a device with a standardized registered number. For example, an HP part 1854,0053 is a (JEDEC) 2N2218 transistor which is also assigned the CV number: CV7763.

Naming Problems

With so many independent naming schemes, and the abbreviation of part numbers when printed on the devices, ambiguity sometimes occurs. For example, two different devices may be marked "J176" (one the J176 low-power JFET, the other the higher-powered MOSFET 2SJ176).

As older "through-hole" transistors are given surface-mount packaged counterparts, they tend to be assigned many different part numbers because manufacturers have their own systems to cope with the variety in pinout arrangements and options for dual or matched n–p–n+p–n–p devices in one pack. So even when the original device (such as a 2N3904) may have been assigned by a standards authority, and well known by engineers over the years, the new versions are far from standardized in their naming.

Construction

Semiconductor Material

Semiconductor material characteristics				
Semiconductor material	**Junction forward voltage V @ 25 °C**	**Electron mobility $m^2/(V \cdot s)$ @ 25 °C**	**Hole mobility $m^2/(V \cdot s)$ @ 25 °C**	**Max. junction temp. °C**
Ge	0.27	0.39	0.19	70 to 100
Si	0.71	0.14	0.05	150 to 200
GaAs	1.03	0.85	0.05	150 to 200
Al-Si junction	0.3	—	—	150 to 200

The first BJTs were made from germanium (Ge). Silicon (Si) types currently predominate but certain advanced microwave and high-performance versions now employ the *compound semiconductor* material gallium arsenide (GaAs) and the *semiconductor alloy* silicon germanium (SiGe). Single element semiconductor material (Ge and Si) is described as *elemental.*

Rough parameters for the most common semiconductor materials used to make transistors are given in the adjacent table; these parameters will vary with increase in temperature, electric field, impurity level, strain, and sundry other factors.

The *junction forward voltage* is the voltage applied to the emitter–base junction of a BJT in order to make the base conduct a specified current. The current increases exponentially as the junction forward voltage is increased. The values given in the table are typical for a current of 1 mA (the same values apply to semiconductor diodes). The lower the junction forward voltage the better, as this means that less power is required to "drive" the transistor. The junction forward voltage for a given current decreases with increase in temperature. For a typical silicon junction the change is −2.1 mV/°C. In some circuits special compensating elements (sensistors) must be used to compensate for such changes.

The density of mobile carriers in the channel of a MOSFET is a function of the electric field forming the channel and of various other phenomena such as the impurity level in the channel. Some impurities, called dopants, are introduced deliberately in making a MOSFET, to control the MOSFET electrical behavior.

The *electron mobility* and *hole mobility* columns show the average speed that electrons and holes diffuse through the semiconductor material with an electric field of 1 volt per meter applied across the material. In general, the higher the electron mobility the faster the transistor can operate. The table indicates that Ge is a better material than Si in this respect. However, Ge has four major shortcomings compared to silicon and gallium arsenide:

- its maximum temperature is limited;
- it has relatively high leakage current;

- it cannot withstand high voltages;
- it is less suitable for fabricating integrated circuits.

Because the electron mobility is higher than the hole mobility for all semiconductor materials, a given bipolar n–p–n transistor tends to be swifter than an equivalent p–n–p transistor. GaAs has the highest electron mobility of the three semiconductors. It is for this reason that GaAs is used in high-frequency applications. A relatively recent FET development, the *high-electron-mobility transistor* (HEMT), has a heterostructure (junction between different semiconductor materials) of aluminium gallium arsenide (AlGaAs)-gallium arsenide (GaAs) which has twice the electron mobility of a GaAs-metal barrier junction. Because of their high speed and low noise, HEMTs are used in satellite receivers working at frequencies around 12 GHz. HEMTs based on gallium nitride and aluminium gallium nitride (AlGaN/GaN HEMTs) provide a still higher electron mobility and are being developed for various applications. Max. junction temperature values represent a cross section taken from various manufacturers' data sheets. This temperature should not be exceeded or the transistor may be damaged. Al–Si junction refers to the high-speed (aluminum–silicon) metal–semiconductor barrier diode, commonly known as a Schottky diode. This is included in the table because some silicon power IGFETs have a *parasitic* reverse Schottky diode formed between the source and drain as part of the fabrication process. This diode can be a nuisance, but sometimes it is used in the circuit.

Packaging

Discrete transistors are individually packaged transistors. Transistors come in many different semiconductor packages (see image). The two main categories are *through-hole* (or *leaded*), and *surface-mount*, also known as *surface-mount device* (SMD). The *ball grid array* (BGA) is the latest surface-mount package (currently only for large integrated circuits). It has solder "balls" on the underside in place of leads. Because they are smaller and have shorter interconnections, SMDs have better high-frequency characteristics but lower power rating.

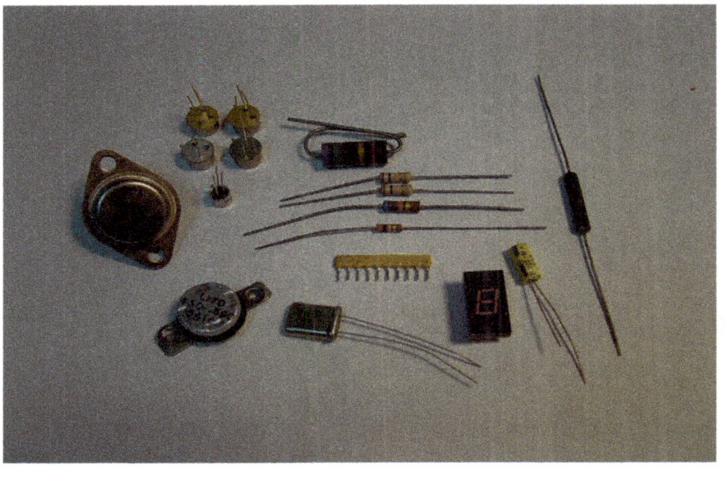

Assorted discrete transistors.

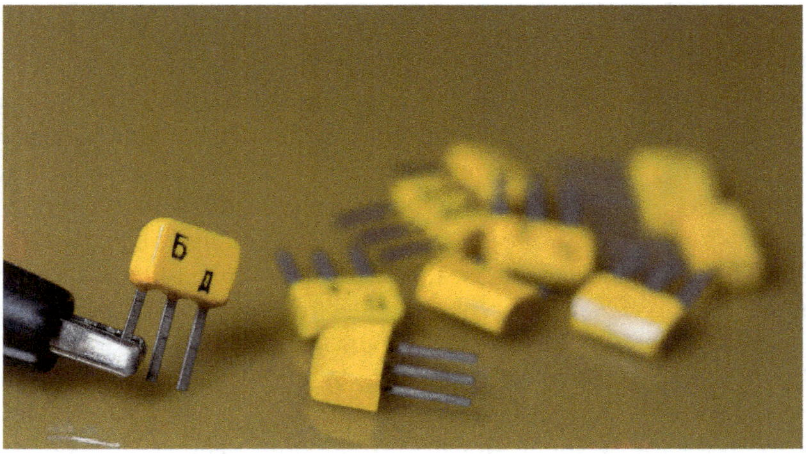

Soviet KT315b transistors.

Transistor packages are made of glass, metal, ceramic, or plastic. The package often dictates the power rating and frequency characteristics. Power transistors have larger packages that can be clamped to heat sinks for enhanced cooling. Additionally, most power transistors have the collector or drain physically connected to the metal enclosure. At the other extreme, some surface-mount *microwave* transistors are as small as grains of sand.

Often a given transistor type is available in several packages. Transistor packages are mainly standardized, but the assignment of a transistor's functions to the terminals is not: other transistor types can assign other functions to the package's terminals. Even for the same transistor type the terminal assignment can vary (normally indicated by a suffix letter to the part number, q.e. BC212L and BC212K).

Nowadays most transistors come in a wide range of SMT packages, in comparison the list of available through-hole packages is relatively small, here is a short list of the most common through-hole transistors packages in alphabetical order: ATV, E-line, MRT, HRT, SC-43, SC-72, TO-3, TO-18, TO-39, TO-92, TO-126, TO220, TO247, TO251, TO262, ZTX851.

Flexible Transistors

Researchers have made several kinds of flexible transistors, including organic field-effect transistors. Flexible transistors are useful in some kinds of flexible displays and other flexible electronics.

TRANSISTOR MODEL

Transistors are simple devices with complicated behavior. In order to ensure the reliable operation of circuits employing transistors, it is necessary to scientifically model the physical phenomena observed in their operation using transistor models. There exists a variety of different models that range in complexity and in purpose. Transistor models divide into two major groups: models for device design and models for circuit design.

Models for Device Design

The modern transistor has an internal structure that exploits complex physical mechanisms. Device design requires a detailed understanding of how device manufacturing processes such as ion implantation, impurity diffusion, oxide growth, annealing, and etching affect device behavior. Process models simulate the manufacturing steps and provide a microscopic description of device "geometry" to the device simulator. "Geometry" does not mean readily identified geometrical features such as a planar or wrap-around gate structure, or raised or recessed forms of source and drain(see Figure for a memory device with some unusual modeling challenges related to charging the floating gate by an avalanche process). It also refers to details inside the structure, such as the doping profiles after completion of device processing.

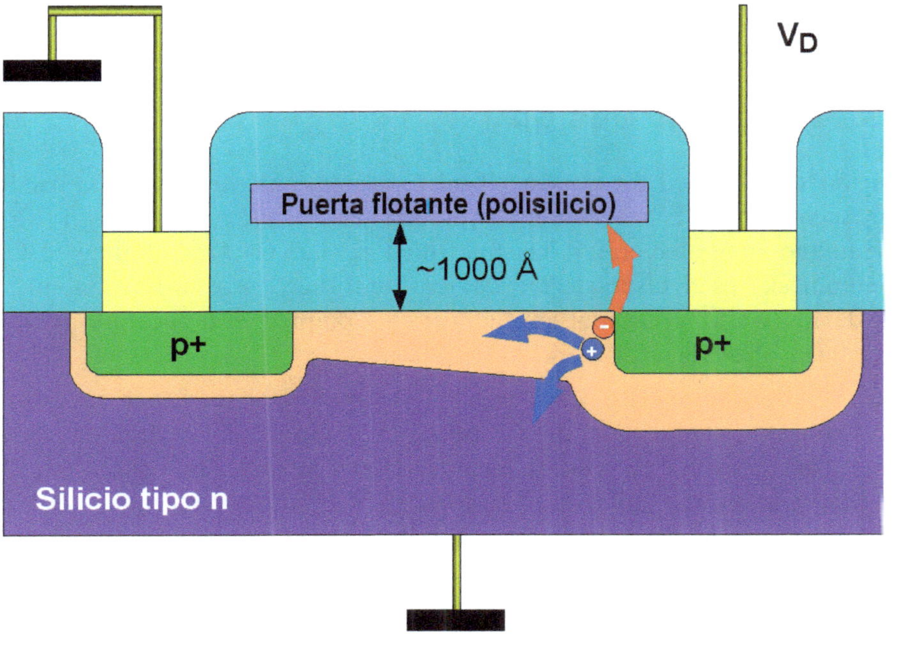

Floating-gate avalanche injection memory device FAMOS

With this information about what the device looks like, the device simulator models the physical processes taking place in the device to determine its electrical behavior in a variety of circumstances: DC current-voltage behavior, transient behavior (both large-signal and small-signal), dependence on device layout (long and narrow versus short and wide, or interdigitated versus rectangular, or isolated versus proximate to other devices). These simulations tell the device designer whether the device process will produce devices with the electrical behavior needed by the circuit designer, and is used to inform the process designer about any necessary process improvements. Once the process gets close to manufacture, the predicted device characteristics are compared with measurement on test devices to check that the process and device models are working adequately.

Although long ago the device behavior modeled in this way was very simple - mainly drift plus diffusion in simple geometries - today many more processes must be modeled at a microscopic level; for example, leakage currents in junctions and oxides, complex transport of carriers including velocity saturation and ballistic transport, quantum mechanical effects, use of multiple materials (for example, Si-SiGe devices, and stacks of different dielectrics) and even the statistical effects due to the probabilistic nature of ion placement and carrier transport inside the device. Several times a year the technology changes and simulations have to be repeated. The models may require change to reflect new physical effects, or to provide greater accuracy. The maintenance and improvement of these models is a business in itself.

These models are very computer intensive, involving detailed spatial and temporal solutions of coupled partial differential equations on three-dimensional grids inside the device. Such models are slow to run and provide detail not needed for circuit design. Therefore, faster transistor models oriented toward circuit parameters are used for circuit design.

Models for Circuit Design

Transistor models are used for almost all modern electronic design work. Analog circuit simulators such as SPICE use models to predict the behavior of a design. Most design work is related to integrated circuit designs which have a very large tooling cost, primarily for the photomasks used to create the devices, and there is a large economic incentive to get the design working without any iterations. Complete and accurate models allow a large percentage of designs to work the first time.

Modern circuits are usually very complex. The performance of such circuits is difficult to predict without accurate computer models, including but not limited to models of the devices used. The device models include effects of transistor layout: width, length, interdigitation, proximity to other devices; transient and DC current-voltage characteristics; parasitic device capacitance, resistance, and inductance; time delays; and temperature effects; to name a few items.

Large-Signal Nonlinear Models

Nonlinear, or large signal transistor models fall into three main types:

Physical Models

These are models based upon device physics, based upon approximate modeling of physical phenomena within a transistor. Parameters within these models are based upon physical properties such as oxide thicknesses, substrate doping concentrations, carrier mobility, etc. In the past these models were used extensively, but the complexity of modern devices makes them inadequate for quantitative design. Nonetheless, they find a place in hand analysis (that is, at the conceptual stage of circuit design), for example, for simplified estimates of signal-swing limitations.

Empirical Models

This type of model is entirely based upon curve fitting, using whatever functions and parameter values most adequately fit measured data to enable simulation of transistor operation. Unlike a physical model, the parameters in an empirical model need have no fundamental basis, and will depend on the fitting procedure used to find them. The fitting procedure is key to success of these models if they are to be used to extrapolate to designs lying outside the range of data to which the models were originally fitted. Such extrapolation is a hope of such models, but is not fully realized so far.

Tabular Models

The third type of model is a form of look-up table containing a large number of values for common device parameters such as drain current and device parasitics. These values are indexed in reference to their corresponding bias voltage combinations. Thus, model accuracy is increased by inclusion of additional data points within the table. The chief advantage of this type of model is decreased simulation time. A limitation of these models is that they work best for designs that use devices within the table (interpolation) and are unreliable for devices outside the table (extrapolation).

The use of nonlinear models, which describe the entire operating area of a transistor, is required for digital designs, for circuits that operate in a large-signal regime such as power amplifiers and mixers, and for the large-signal simulation of any circuit, for example, for stability or distortion analysis.

Nonlinear models are used with a computer simulation program, such as SPICE. The models in SPICE are a hybrid of physical and empirical models, and such models are incomplete unless they include specification of how parameter values are to be extracted, especially as "unrealistic" (that is, unphysical) values can be made to fit the measured data without such a prescription. An incorrect set of fitting parameters results in wild predictions for devices that were not part of the originally fitted data set.

Large-signal computer models for devices continually evolve to keep up with changes in technology. To attempt standardization of model parameters used in different simulators, an industry working group was formed, the Compact Model Council, to choose, maintain and promote the use of standard models. An elusive goal in such modeling is prediction of how circuits using the next generation of devices should work, to identify before the next step which direction the technology should take, and have models ready beforehand.

Small-Signal Linear Models

Small-signal or linear models are used to evaluate stability, gain, noise and bandwidth, both in the conceptual stages of circuit design (to decide between alternative design ideas before computer simulation is warranted) and using computers. A small-signal model is generated by taking derivatives of the current-voltage curves about a bias point or Q-point. As long as the signal is small relative to the nonlinearity of the device, the derivatives do not vary significantly, and can be treated as standard linear circuit elements. A big advantage of small signal models is they can be solved directly, while large signal nonlinear models are generally solved iteratively, with possible convergence or stability issues. By simplification to a linear model, the whole apparatus for solving linear equations becomes available, for example, simultaneous equations, determinants, and matrix theory (often studied as part of linear algebra), especially Cramer's rule. Another advantage is that a linear model is easier to think about, and helps to organize thought.

Small-Signal Parameters

A transistor's parameters represent its electrical properties. Engineers employ transistor parameters in production-line testing and in circuit design. A group of a transistor's parameters sufficient to predict circuit gain, input impedance, and output impedance are components in its small-signal model.

A number of different two-port network parameter sets may be used to model a transistor. These include:

- Transmission parameters (T-parameters),
- Hybrid-parameters (h-parameters),
- Impedance parameters (z-parameters),
- Admittance parameters (y-parameters), and
- Scattering parameters (S-parameters).

Scattering parameters, or S parameters, can be measured for a transistor at a given bias point with a vector network analyzer. S parameters can be converted to another parameter set using standard matrix algebra operations.

TRANSISTOR COUNT

The transistor count is the number of transistors on an integrated circuit (IC). Transistor count is the most common measure of IC complexity, although there are caveats. For instance, the majority of transistors are contained in the cache memories in modern microprocessors, which consist mostly of the same memory cell circuits replicated many times. The rate at which transistor counts have increased generally follows Moore's law, which observed that the transistor count doubles approximately every two years. As of 2016, the largest transistor count in a commercially available single-chip processor is over 7.2 billion—the Intel Broadwell-EP Xeon. In other types of ICs, such as field-programmable gate arrays (FPGAs), Intel's (previously Altera) Stratix 10 has the largest transistor count, containing over 30 billion transistors.

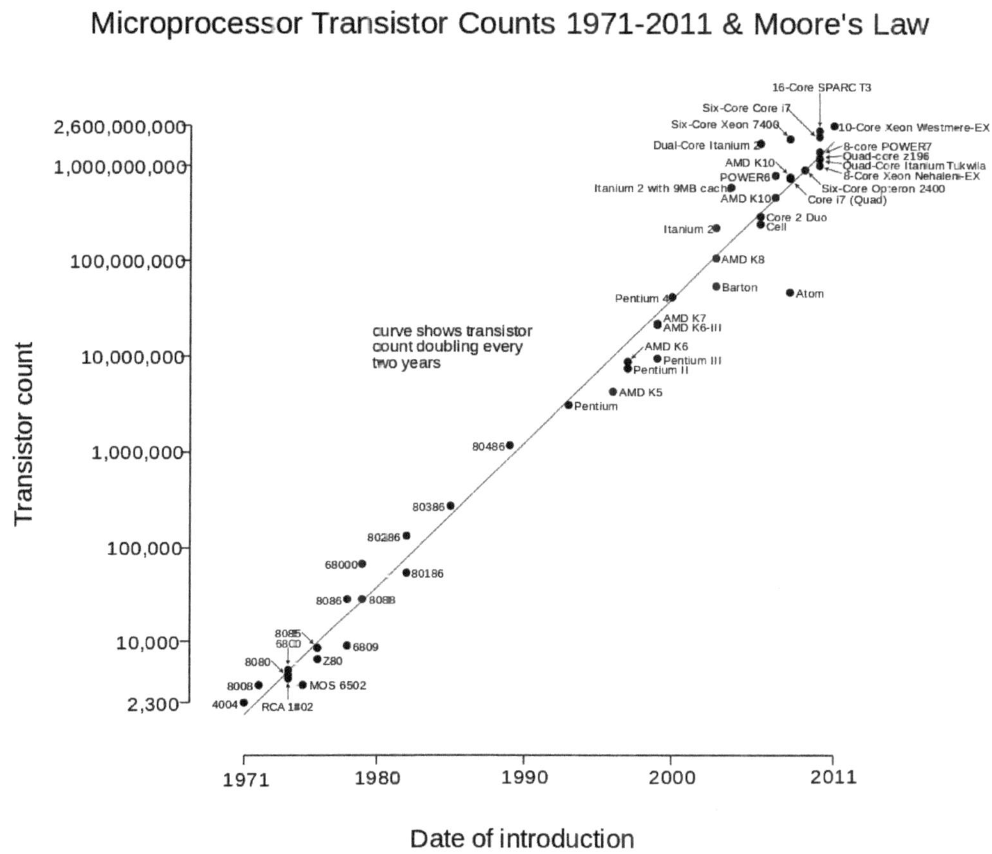

Plot of **transistor counts** against dates of introduction. The curve shows counts doubling every two years, per Moore's law.

Microprocessors

A microprocessor incorporates the functions of a computer's central processing unit on a single integrated circuit. It is a multipurpose, programmable device that accepts digital data as input, processes it according to instructions stored in its memory, and provides results as output.

Processor	Transistor count	Date of introduction	Designer	Process	Area
TMS 1000	8,000	1974	Texas Instruments	8,000 nm	
Intel 4004	2,300	1971	Intel	10,000 nm	12 mm²
Intel 8008	3,500	1972	Intel	10,000 nm	14 mm²
MOS Technology 6502	3,510	1975	MOS Technology	8,000 nm	21 mm²

Motorola 6800	4,100	1974	Motorola	6,000 nm	16 mm^2
Intel 8080	4,500	1974	Intel	6,000 nm	20 mm^2
RCA 1802	5,000	1974	RCA	5,000 nm	27 mm^2
Intel 8085	6,500	1976	Intel	3,000 nm	20 mm^2
Zilog Z80	8,500	1976	Zilog	4,000 nm	18 mm^2
Motorola 6809	9,000	1978	Motorola	5,000 nm	21 mm^2
Intel 8086	29,000	1978	Intel	3,000 nm	33 mm^2
Intel 8088	29,000	1979	Intel	3,000 nm	33 mm^2
WDC 65C02	11,500	1981	WDC	3,000 nm	6 mm^2
Intel 80186	55,000	1982	Intel	3,000 nm	60 mm^2
Motorola 68000	68,000	1979	Motorola	3,500 nm	44 mm^2
Intel 80286	134,000	1982	Intel	1,500 nm	49 mm^2
WDC 65C816	22,000	1983	WDC		9 mm^2
Motorola 68020	190,000	1984	Motorola	2,000 nm	85 mm^2
Intel 80386	275,000	1985	Intel	1,500 nm	104 mm^2
ARM 1	25,000	1985	Acorn	3,000 nm	50 mm^2
Novix NC4016	16,000	1985	Harris Corporation	3,000 nm	
ARM 2	30,000	1986	Acorn	2,000 nm	30 mm^2
TI Explorer's 32-bit Lisp machine chip	553,000	1987	Texas Instruments		
DEC WRL MultiTitan	180,000	1988	DEC WRL	1,500 nm	61 mm^2
Intel i960	250,000	1988	Intel	600 nm	
Intel 80486	1,180,235	1989	Intel	1000 nm	173 mm^2
ARM 3	300,000	1989	Acorn		
R4000	1,350,000	1991	MIPS	1,000 nm	213 mm^2
ARM 6	35,000	1991	ARM		
Pentium	3,100,000	1993	Intel	800 nm	294 mm^2
ARM700	578,977	1994	ARM		68.51 mm^2
SA-110	2,500,000	1995	Acorn/DEC/Apple	350 nm	50 mm^2
ARM 9TDMI	111,000	1999	Acorn	350 nm	4.8 mm^2
Pentium Pro	5,500,000	1995	Intel	500 nm	307 mm^2
AMD K5	4,300,000	1996	AMD	500 nm	251 mm^2
Pentium II Klamath	7,500,000	1997	Intel	350 nm	195 mm^2
Pentium II Deschutes	7,500,000	1998	Intel	250 nm	113 mm^2
AMD K6	8,800,000	1997	AMD	350 nm	162 mm^2
Pentium III Katmai	9,500,000	1999	Intel	250 nm	128 mm^2
Pentium III Coppermine	21,000,000	2000	Intel	180 nm	80 mm^2
Pentium II Mobile Dixon	27,400,000	1999	Intel	180 nm	180 mm^2
Pentium III Tualatin	45,000,000	2001	Intel	130 nm	81 mm^2
AMD K6-III	21,300,000	1999	AMD	250 nm	118 mm^2

AMD K7	22,000,000	1999	AMD	250 nm	184 mm^2
Pentium 4 Willamette	42,000,000	2000	Intel	180 nm	217 mm^2
Pentium 4 Northwood	55,000,000	2002	Intel	130 nm	145 mm^2
Pentium 4 Prescott	112,000,000	2004	Intel	90 nm	110 mm^2
Pentium 4 Prescott-2M	169,000,000	2005	Intel	90 nm	143 mm^2
Pentium 4 Cedar Mill	184,000,000	2006	Intel	65 nm	90 mm^2
Pentium D Smithfield	228,000,000	2005	Intel	90 nm	206 mm^2
Pentium D Presler	362,000,000	2006	Intel	65 nm	162 mm^2
Atom	47,000,000	2008	Intel	45 nm	24 mm^2
Barton	54,300,000	2003	AMD	130 nm	101 mm^2
AMD K8	105,900,000	2003	AMD	130 nm	193 mm^2
Itanium 2 McKinley	220,000,000	2002	Intel	180 nm	421 mm^2
Cell	241,000,000	2006	Sony/IBM/ Toshiba	90 nm	221 mm^2
Core 2 Duo Conroe	291,000,000	2006	Intel	65 nm	143 mm^2
Core 2 Duo Allendale	169,000,000	2007	Intel	65 nm	111 mm^2
Itanium 2 Madison 6M	410,000,000	2003	Intel	130 nm	374 mm^2
AMD K10 quad-core 2M L3	463,000,000	2007	AMD	65 nm	283 mm^2
ARM Cortex-A9	26,000,000	2007	ARM	45 nm	31 mm^2
Core 2 Duo Wolfdale 3M	230,000,000	2008	Intel	45 nm	83 mm^2
Itanium 2 with 9 MB cache	592,000,000	2004	Intel	130 nm	432 mm^2
Core 2 Duo Wolfdale	411,000,000	2007	Intel	45 nm	107 mm^2
Core i7 (Quad)	731,000,000	2008	Intel	45 nm	263 mm^2
AMD K10 quad-core 6M L3	758,000,000	2008	AMD	45 nm	258 mm^2
POWER6	789,000,000	2007	IBM	65 nm	341 mm^2
Six-core Opteron 2400	904,000,000	2009	AMD	45 nm	346 mm^2
16-core SPARC T3	1,000,000,000	2010	Sun/Oracle	40 nm	377 mm^2
Apple A7 (dual-core ARM64 "mobile SoC")	1,000,000,000	2013	Apple	28 nm	102 mm^2
Quad-core + GPU Core i7	1,160,000,000	2011	Intel	32 nm	216 mm^2
Six-core Core i7 (Gulftown)	1,170,000,000	2010	Intel	32 nm	240 mm^2
8-core POWER7 32M L3	1,200,000,000	2010	IBM	45 nm	567 mm^2
8-core AMD Bulldozer	1,200,000,000	2012	AMD	32 nm	315 mm^2
Quad-core + GPU AMD Trinity	1,303,000,000	2012	AMD	32 nm	246 mm^2
Quad-core z196	1,400,000,000	2010	IBM	45 nm	512 mm^2
Quad-core + GPU Core i7 Ivy Bridge	1,400,000,000	2012	Intel	22 nm	160 mm^2

Quad-core + GPU Core i7 Haswell	1,400,000,000	2014	Intel	22 nm	177 mm^2
Dual-core Itanium 2	1,700,000,000	2006	Intel	90 nm	596 mm^2
Quad-core + GPU GT2 Core i7 Skylake K	1,750,000,000	2015	Intel	14 nm	122 mm^2
Six-core Core i7 Ivy Bridge E	1,860,000,000	2013	Intel	22 nm	256 mm^2
Duo-core + GPU Iris Core i7 Broadwell-U	1,900,000,000	2015	Intel	14 nm	133 mm^2
Six-core Xeon 7400	1,900,000,000	2008	Intel	45 nm	503 mm^2
Quad-core Itanium Tukwila	2,000,000,000	2010	Intel	65 nm	699 mm^2
Apple A8 (dual-core ARM64 "mobile SoC")	2,000,000,000	2014	Apple	20 nm	89 mm^2
8-core POWER7+ 80 MB L3 cache	2,100,000,000	2012	IBM	32 nm	567 mm^2
Six-core Core i7/8-core Xeon E5 (Sandy Bridge-E/EP)	2,270,000,000	2011	Intel	32 nm	434 mm^2
8-core Xeon Nehalem-EX	2,300,000,000	2010	Intel	45 nm	684 mm^2
8-core Core i7 Haswell-E	2,600,000,000	2014	Intel	22 nm	355 mm^2
10-core Xeon Westmere-EX	2,600,000,000	2011	Intel	32 nm	512 mm^2
Six-core zEC12	2,750,000,000	2012	IBM	32 nm	597 mm^2
Apple A8X (tri-core ARM64 "mobile SoC")	3,000,000,000	2014	Apple	20 nm	128 mm^2
8-core Itanium Poulson	3,100,000,000	2012	Intel	32 nm	544 mm^2
Apple A10 (quad-core ARM64 "mobile SoC")	3,300,000,000	2016	Apple	16 FFC	125 mm^2
IBM z13	3,990,000,000	2015	IBM	22 nm	678 mm^2
12-core POWER8	4,200,000,000	2013	IBM	22 nm	650 mm^2
15-core Xeon Ivy Bridge-EX	4,310,000,000	2014	Intel	22 nm	541 mm^2
61-core Xeon Phi	5,000,000,000	2012	Intel	22 nm	720 mm^2
Xbox One main SoC	5,000,000,000	2013	Microsoft/AMD	28 nm	363 mm^2
18-core Xeon Haswell-E5	5,560,000,000	2014	Intel	22 nm	661 mm^2
IBM z13 Storage Controller	7,100,000,000	2015	IBM	22 nm	678 mm^2
22-core Xeon Broadwell-E5	7,200,000,000	2016	Intel	14 nm	456 mm^2
SPARC M7	10,000,000,000	2015	Oracle	20 nm	

Transistorized Computers

The "second generation" of computers (transistor computers) featured boards filled with discrete transistors and magnetic memory cores.

Processor	Transistor count	Date of introduction	Manufacturer	Process	Area
"Transistor Computer" (full size)	200 discrete point-contact transistors	1955	University of Manchester	?	?
"Metrovick 950"	200 discrete junction transistors	1956	Metropolitan-Vickers	?	?
IBM 7090	50,000+ discrete germanium transistors	1959	IBM		
PDP-1	2,700 discrete transistors	1959	Digital Equipment Corporation		
M18 FADAC	1,600 discrete transistors	1960	Autonetics	?	?
D-17B	1,521 discrete transistors	1962	Autonetics	?	?

Early Integrated Circuit Computers

Computer	Manufacturer	Year	Transistors	Notes
PDP-8/s	Digital Equipment Corporation	1971?	?	519 logic gates
Apollo Guidance Computer Block I	Raytheon MIT Instrumentation Laboratory	1966	12,300	4,100 ICs, each containing a 3-transistor, 3-input NOR gate

GPUs

A graphics processing unit (GPU) is a specialized electronic circuit designed to rapidly manipulate and alter memory to accelerate the building of images in a frame buffer intended for output to a display.

Processor	Transistor count	Date of introduction	Manufacturer	Process	Area
NV3	3,500,000	1997	NVIDIA	350 nm	90 mm²
Rage 128	8,000,000	1999	AMD	250 nm	70 mm²
NV5	15,000,000	1999	Nvidia	250 nm	
NV10	23,000,000	1999	Nvidia	220 nm	111 mm²
NV11	20,000,000	2000	Nvidia	180 nm	65 mm²
NV15	25,000,000	2000	Nvidia	180 nm	81 mm²
R100	30,000,000	2000	AMD	180 nm	97 mm²

NV20	57,000,000	2001	Nvidia	150 nm	128 mm²
R200	60,000,000	2001	AMD	150 nm	68 mm²
NV25	63,000,000	2002	Nvidia	150 nm	142 mm²
R300	107,000,000	2002	AMD	150 nm	218 mm²
R360	117,000,000	2003	AMD	150 nm	218 mm²
NV38	135,000,000	2003	Nvidia	130 nm	207 mm²
R480	160,000,000	2004	AMD	130 nm	297 mm²
NV40	222,000,000	2004	Nvidia	130 nm	305 mm²
G70	303,000,000	2005	Nvidia	110 nm	333 mm²
R520	321,000,000	2005	AMD	90 nm	288 mm²
R580	384,000,000	2006	AMD	90 nm	352 mm²
G80	681,000,000	2006	Nvidia	90 nm	480 mm²
R600	700,000,000	2007	AMD	80 nm	420 mm²
G92	754,000,000	2007	Nvidia	65 nm	324 mm²
RV790	959,000,000	2008	AMD	55 nm	282 mm²
GT200 Tesla	1,400,000,000	2008	Nvidia	65 nm	576 mm²
Cypress RV870	2,154,000,000	2009	AMD	40 nm	334 mm²
Cayman RV970	2,640,000,000	2010	AMD	40 nm	389 mm²
GF100 Fermi	3,200,000,000	Mar 2010	Nvidia	40 nm	526 mm²
GF110 Fermi	3,000,000,000	Nov 2010	Nvidia	40 nm	520 mm²
GK104 Kepler	3,540,000,000	2012	Nvidia	28 nm	294 mm²
Tahiti	4,312,711,873	2011	AMD	28 nm	365 mm²
GK110 Kepler	7,080,000,000	2012	Nvidia	28 nm	561 mm²
Hawaii	6,300,000,000	2013	AMD	28 nm	438 mm²
GM204 Maxwell	5,200,000,000	2014	Nvidia	28 nm	398 mm²
GM200 Maxwell	8,100,000,000	2015	Nvidia	28 nm	601 mm²
Fiji	8,900,000,000	2015	AMD	28 nm	596 mm²
GP104 Pascal	7,200,000,000	2016	Nvidia	16 nm	314 mm²
GP100 Pascal	15,300,000,000	2016	Nvidia	16 nm	610 mm²

FPGA

A field-programmable gate array (FPGA) is an integrated circuit designed to be configured by a customer or a designer after manufacturing.

FPGA	Transistor count	Date of introduction	Manufacturer	Process	Area
Virtex	~70,000,000	1997	Xilinx		
Virtex-E	~200,000,000	1998	Xilinx		
Virtex-II	~350,000,000	2000	Xilinx	130 nm	
Virtex-II PRO	~430,000,000	2002	Xilinx		
Virtex-4	1,000,000,000	2004	Xilinx	90 nm	
Virtex-5	1,100,000,000	2006	Xilinx	65 nm	
Stratix IV	2,500,000,000	2008	Altera	40 nm	
Stratix V	3,800,000,000	2011	Altera	28 nm	
Virtex-7	6,800,000,000	2011	Xilinx	28 nm	
Virtex-Ultrascale XCVU440	20,000,000,000+	2014	Xilinx	20 nm	
Stratix 10 Family device, 10GX5500/10SX5500	30,000,000,000+ *	2016	Intel (formally Altera)	14 nm	

Logic Functions

Transistor count for generic logic functions is based on static CMOS implementation.

Function	Transistor count
NOT	2
Buffer	4
NAND 2-input	4
NOR 2-input	4
AND 2-input	6
OR 2-input	6
NAND 3-input	6
NOR 3-input	6
XOR 2-input	6
XNOR 2-input	8
MUX 2-input with TG	6
MUX 4-input with TG	18
NOT MUX 2-input	8
MUX 4-input	24
1-bit Adder full	28
1-bit Adder–subtractor	48

AND-OR-INVERT	6
Latch, D gated	8
Flip-flop, edge triggered dynamic D with reset	12
8-bit multiplier	3,000
16-bit multiplier	9,000
32-bit multiplier	21,000
small-scale integration	2–100
medium-scale integration	100–500
large-scale integration	500–20,000
very-large-scale integration	20,000–1,000,000
ultra-large scale integration	>1,000,000

Memory

Semiconductor memory is an electronic data storage device, often used as computer memory, implemented on integrated circuit.

We know that in order to store a single bit (which may be 1 or 0), one flip-flop is required, made of around eight transistors. Typical CMOS Static random-access memory (SRAM) consists of 6 transistors. For Dynamic random-access memory (DRAM), 1T1C, which means one transistor and one capacitor structure is common. Capacitor charged or not is used to store 1 or 0. For flash memory, the data is stored in floating gate, and the resistance of the transistor is sensed to interpret the data stored. Depending on how fine scale the resistance could be separated, one transistor could store up to 3-bits, meaning eight distinctive level of resistance possible per transistor. However, the fine the scale comes with cost of repeatability therefore reliability. Typically, low grade 2-bits MLC flash is used for flash drive, so a 16 GB flash drive contains roughly 64 billion transistors.

Chip	**Capacity & type**	**Transistor count**	**Date of introduction**	**Manufacturer**	**Process**	**Area**
?	256-bit ROM bipolar TTL	?	1965	Sylvania	?	?
?	1024-bit ROM MOS	?	1965	General Microelectronics	?	?
SP95	16-bit SRAM bipolar	?	1965	IBM	?	?
?	128-bit RAM	?	1969	IBM	?	?
	512-bit PROM bipolar TTL	?	1970	Radiation Inc.	?	?
93400	256-bit RAM	?	1970	Fairchild	?	?
1103	1 kb DRAM	?1,024	1970	Intel	?	?
1702 Erasable PROM	2 kb EPROM	?	1971	Intel	?	?
?	8 Mb DRAM	?8,388,608	January 6, 1984 (1986)	Hitachi	?	?

?	64 Mb DRAM	?67,108,864	1994	NEC, Samsung	320 nm	?
?	256 Mb DRAM	?268,435,456	June 12, 1995	IBM, SIEMENS AG, Toshiba Corp.	250 nm	286 mm^2
?	1 Gb DRAM	?1,073,741,824	January 9, 1995 (2001)	Hitachi	?	?
	4 Gb DRAM					
	8 Gb DRAM					
	16 Gb DRAM					
	32 Gb DRAM					
?	64 Gb DRAM	?68,719,476,736	2007(9)?	?	?	?
?	128 Gb DRAM	?137,438,953,472	July 5, 2012	Samsung	30 nm	?

Parallel Systems

Historically, each processing element in earlier parallel systems—like all CPUs of that time—was a serial computer built out of multiple chips. As transistor counts per chip increases, each processing element could be built out of fewer chips, and then later each multi-core processor chip could contain more processing elements.

- Goodyear MPP : (1983?) 8 pixel processors per chip, 3,000 to 8,000 transistors per chip.
- Brunel University Scape (single-chip array-processing element): (1983) 256 pixel processors per chip, 120,000 to 140,000 transistors per chip.
- Cell Broadband Engine: (2006) 9 cores per chip, 234 million transistors per chip.

REFERENCES

1. "1926 – Field Effect Semiconductor Device Concepts Patented". Computer history museum. Retrieved March 25, 2016.
2. "Milestones:Invention of the First Transistor at Bell Telephone Laboratories, Inc., 1947". IEEE Global History Network. IEEE. Retrieved December 7, 2014.
3. "The Nobel Prize in Physics 1956". Nobelprize.org. Nobel Media AB. Retrieved December 7, 2014.
4. Vardalas, John (May 2003) Twists and Turns in the Development of the Transistor IEEE-USA Today's Engineer.
5. Lilienfeld, Julius Edgar, "Method and apparatus for controlling electric current" U.S. Patent 1,745,175 January 28, 1930 (filed in Canada 1925-10-22, in US October 8, 1926).
6. "Method And Apparatus For Controlling Electric Currents". United States Patent and Trademark Office.
7. "Amplifier For Electric Currents". United States Patent and Trademark Office.
8. "Device For Controlling Electric Current". United States Patent and Trademark

Office.

9. "Twists and Turns in the Development of the Transistor". Institute of Electrical and Electronics Engineers, Inc.
10. Heil, Oskar, "Improvements in or relating to electrical amplifiers and other control arrangements and devices", Patent No. GB439457, European Patent Office, filed in Great Britain 1934-03-02, published December 6, 1935 (originally filed in Germany March 2, 1934).
11. "November 17 – December 23, 1947: Invention of the First Transistor". American Physical Society.
12. Millman, S., ed. (1983). A History of Engineering and Science in the Bell System, Physical Science (1925–1980). AT&T Bell Laboratories. p. 102.
13. Bodanis, David (2005). Electric Universe. Crown Publishers, New York. ISBN 0-7394-5670-9.
14. "transistor". American Heritage Dictionary (3rd ed.). Boston: Houghton Mifflin. 1992.
15. "The Nobel Prize in Physics 1956". nobelprize.org.
16. "1948, The European Transistor Invention". Computer History Museum.
17. Bradley, W.E. (December 1953). "The Surface-Barrier Transistor: Part I-Principles of the Surface-Barrier Transistor". Proceedings of the IRE. 41 (12): 1702–1706. doi:10.1109/JRPROC.1953.274351.
18. Wall Street Journal, December 4, 1953, page 4, Article "Philco Claims Its Transistor Outperforms Others Now In Use"
19. Electronics magazine, January 1954, Article "Electroplated Transistors Announced"
20. "Der deutsche Erfinder des Transistors – Nachrichten Welt Print – DIE WELT". Welt.de. November 23, 2011. Retrieved May 1, 2016.
21. Wall Street Journal, "Chrysler Promises Car Radio With Transistors Instead of Tubes in '56", April 28, 1955, page 1
22. Los Angeles Times, May 8, 1955, page A20, Article: "Chrysler Announces New Transistor Radio"
23. Philco TechRep Division Bulletin, May–June 1955, Volume 5 Number 3, page 28
24. Rosen, Saul (Jun 1991). PHILCO: Some Recollections of the PHILCO TRANSAC S-2000 (Computer Science Technical Reports / Purdue e-Pubs). Purdue University. p. 2.
25. Riordan, Michael (May 2004). "The Lost History of the Transistor". IEEE Spectrum: 48–49.
26. Chelikowski, J. (2004) "Introduction: Silicon in all its Forms", p. 1 in Silicon: evolution and future of a technology. P. Siffert and E. F. Krimmel (eds.). Springer, ISBN 3-540-40546-1.
27. McFarland, Grant (2006) Microprocessor design: a practical guide from design planning to manufacturing. McGraw-Hill Professional. p. 10. ISBN 0-07-145951-0.

28. Heywang, W. and Zaininger, K. H. (2004) "Silicon: The Semiconductor Material", p. 36 in Silicon: evolution and future of a technology. P. Siffert and E. F. Krimmel (eds.). Springer, 2004 ISBN 3-540-40546-1.
29. Price, Robert W. (2004). Roadmap to Entrepreneurial Success. AMACOM Div American Mgmt Assn. p. 42. ISBN 978-0-8144-7190-6.
30. "Milestones:Invention of the First Transistor at Bell Telephone Laboratories, Inc., 1947". IEEE Global History Network. IEEE. Retrieved August 3, 2011.
31. FETs/MOSFETs: Smaller apps push up surface-mount supply. globalsources.com (April 18, 2007)
32. "ATI and Nvidia face off." CNET (October 7, 2009). Retrieved on February 2, 2011.
33. Turley, Jim (December 18, 2002). "The Two Percent Solution". embedded.com
34. Kaplan, Daniel (2003). Hands-On Electronics. New York: Cambridge University Press. pp. 47–54, 60–61. ISBN 978-0-511-07668-8.
35. John Keane and Chris H. Kim, "Transistor Aging," IEEE Spectrum (web feature), April 25, 2011.
36. van der Veen, M. (2005). "Universal system and output transformer for valve amplifiers" (PDF). 118th AES Convention, Barcelona, Spain.
37. "Transistor Example". 071003 bcae1.com
38. Streetman, Ben (1992). Solid State Electronic Devices. Englewood Cliffs, NJ: Prentice-Hall. pp. 301–305. ISBN 0-13-822023-9.
39. Horowitz, Paul; Winfield Hill (1989). The Art of Electronics (2nd ed.). Cambridge University Press. p. 115. ISBN 0-521-37095-7.
40. Sansen, W. M. C. (2006). Analog design essentials. New York, Berlin: Springer. p. §0152, p. 28. ISBN 0-387-25746-2.
41. IGBT Module 5SNA 2400E170100 (PDF). Retrieved June 30, 2012.
42. Zhong Yuan Chang, Willy M. C. Sansen, Low-Noise Wide-Band Amplifiers in Bipolar and CMOS Technologies, page 31, Springer, 1991 ISBN 0792390962.
43. Single Electron Transistors. Snow.stanford.edu. Retrieved June 30, 2012.
44. Sanders, Robert (June 28, 2005). Nanofluidic transistor, the basis of future chemical processors. Berkeley.edu. Retrieved June 30, 2012.
45. The return of the vacuum tube?. Gizmag.com. May 28, 2012. Retrieved May 1, 2016.
46. Transistor Data. Clivetec.0catch.com. Retrieved May 1, 2016.
47. Datasheet for BC549, with A,B and C gain groupings (PDF). Fairchild Semiconductor. Retrieved June 30, 2012.
48. Datasheet for BUK854-800A (800volt IGBT) (PDF). Retrieved June 30, 2012.
49. Richard Freeman›s HP Part numbers Crossreference. Hpmuseum.org. Retrieved June 30, 2012.

50. Transistor–Diode Cross Reference – H.P. Part Numbers to JEDEC (pdf) (PDF). Retrieved May 1, 2016.

51. CV Device Cross-reference by Andy Lake. Qsl.net. Retrieved June 30, 2012.

52. Sedra, A.S. & Smith, K.C. (2004). Microelectronic circuits (Fifth ed.). New York: Oxford University Press. p. 397 and Figure 5.17. ISBN 0-19-514251-9.

53. Rojas, Jhonathan P.; Torres Sevilla, Galo A.; Hussain, Muhammad M. (2013). Can We Build a Truly High Performance Computer Which is Flexible and Transparent?. Scientific Reports. 3. Bibcode:2013NatSR...3E2609R. doi:10.1038/srep02609.

54. Zhang, Kan; Seo, Jung-Hun; Zhou, Weidong; Ma, Zhenqiang (2012). Fast flexible electronics using transferrable silicon nanomembranes. Journal of Physics D: Applied Physics. 45 (14): 143001. Bibcode:2012JPhD...45n3001Z. doi:10.1088/0022-3727/45/14/143001.

55. Sun, Dong-Ming; Timmermans, Marina Y.; Tian, Ying; Nasibulin, Albert G.; Kauppinen, Esko I.; Kishimoto, Shigeru; Mizutani, Takashi; Ohno, Yutaka (2011). Flexible high-performance carbon nanotube integrated circuits. Nature Nanotechnology. 6 (3): 156. Bibcode:2011NatNa...6..156S. doi:10.1038/NNANO.2011.1. PMID 21297625.

CHAPTER 3

Capacitors

Chapter Contents

- **Ntroduction**
- **History**
- **Theory of Operation**
- **Non-Ideal Behavior**
- **Capacitor Markings**
- **Applications**
- **Capacitance Meter**
- **Capacitor Plague**

The capacitor is a component which has the ability or "capacity" to store energy in the form of an electrical charge producing a potential difference (Static Voltage) across its plates, much like a small rechargeable battery.

There are many different kinds of capacitors available from very small capacitor beads used in resonance circuits to large power factor correction capacitors, but they all do the same thing, they store charge.

In its basic form, a capacitor consists of two or more parallel conductive (metal) plates which are not connected or touching each other, but are electrically separated either by air or by some form of a good insulating material such as waxed paper, mica, ceramic, plastic or some form of a liquid gel as used in electrolytic capacitors. The insulating layer between a capacitors plates is commonly called the Dielectric.

INTRODUCTION

A capacitor is a passive two-terminal electrical component that stores electrical energy in an electric field. The effect of a capacitor is known as capacitance. While capacitance exists between any two electrical conductors of a circuit in sufficiently close proximity, a capacitor is specifically designed to provide and enhance this effect for a variety of practical applications by consideration of size, shape, and positioning of closely spaced conductors, and the intervening dielectric material. A capacitor was therefore historically first known as an electric condenser.

The physical form and construction of practical capacitors vary widely and many capacitor types are in common use. Most capacitors contain at least two electrical conductors often in the form of metallic plates or surfaces separated by a dielectric medium. The conductors may be foils, thin films, or sintered beads of metal or conductive electrolyte. The nonconducting dielectric acts to increase the capacitor's charge capacity. Materials commonly used as dielectrics include glass, ceramic, plastic film, paper, mica, and oxide layers. Capacitors are widely used as parts of electrical circuits in many common electrical devices. Unlike a resistor, an ideal capacitor does not dissipate energy.

When two conductors experience a potential difference, for example, when a capacitor is attached across a battery, an electric field develops across the dielectric, causing a net positive charge to collect on one plate and net negative charge to collect on the other plate. No current actually flows through the dielectric, instead, the effect is a displacement of charges through the source circuit. If the condition is maintained sufficiently long, this displacement current through the battery ceases. However, if a time-varying voltage is applied across the leads of the capacitor, the source experiences an ongoing current due to the charging and discharging cycles of the capacitor.

Capacitance is defined as the ratio of the electric charge Q on each conductor to the potential difference V between them. The unit of capacitance in the International System of Units (SI) is the farad (F), which is equal to one coulomb per volt (1 C/V). Capacitance values of typical capacitors for use in general electronics range from about 1 pF (10^{-12} F) to about 1 mF (10^{-3} F).

The capacitance of a capacitor is proportional to the surface area of the plates (conductors) and inversely related to the gap between them. In practice, the dielectric between the plates passes a small amount of leakage current and also has an electric field strength limit, known as the breakdown voltage. The conductors and leads introduce an undesired inductance and resistance.

Capacitors are widely used in electronic circuits for blocking direct current while allowing alternating current to pass. In analog filter networks, they smooth the output of power supplies. In resonant circuits they tune radios to particular frequencies. In electric power transmission systems, they stabilize voltage and power flow. The property of energy storage in capacitors was exploited as dynamic memory in early digital computers.

HISTORY

In October 1745, Ewald Georg von Kleist of Pomerania, Germany, found that charge could be stored by connecting a high-voltage electrostatic generator by a wire to a volume of water in a hand-held glass jar. Von Kleist's hand and the water acted as conductors, and the jar as a dielectric (although details of the mechanism were incorrectly identified at the time). Von Kleist found that touching the wire resulted in a powerful spark, much more painful than that obtained from an electrostatic machine. The following year, the

Dutch physicist Pieter van Musschenbroek invented a similar capacitor, which was named the Leyden jar, after the University of Leiden where he worked. He also was impressed by the power of the shock he received, writing, "I would not take a second shock for the kingdom of France."

Battery of four Leyden jars in Museum Boerhaave, Leiden, the Netherlands

Daniel Gralath was the first to combine several jars in parallel into a "battery" to increase the charge storage capacity. Benjamin Franklin investigated the Leyden jar and came to the conclusion that the charge was stored on the glass, not in the water as others had assumed. He also adopted the term "battery", (denoting the increasing of power with a row of similar units as in a battery of cannon), subsequently applied to clusters of electrochemical cells. Leyden jars were later made by coating the inside and outside of jars with metal foil, leaving a space at the mouth to prevent arcing between the foils. The earliest unit of capacitance was the jar, equivalent to about 1.11 nanofarads.

Leyden jars or more powerful devices employing flat glass plates alternating with foil conductors were used exclusively up until about 1900, when the invention of wireless (radio) created a demand for standard capacitors, and the steady move to higher frequencies required capacitors with lower inductance. More compact construction methods began to be used, such as a flexible dielectric sheet (like oiled paper) sandwiched between sheets of metal foil, rolled or folded into a small package.

Early capacitors were known as condensers, a term that is still occasionally used today, particularly in high power applications, such as automotive systems. The term was first used for this purpose by Alessandro Volta in 1782, with reference to the device's ability to store a higher density of electric charge than was possible with an isolated conductor. The term became deprecated because of the ambiguous meaning of steam condenser, with capacitor becoming the recommended term from 1926.

Since the beginning of the study of electricity non conductive materials like glass, porcelain, paper and mica have been used as insulators. These materials some decades later were also well-suited for further use as the dielectric for the first capacitors. Paper capacitors made by sandwiching a strip of impregnated paper between strips of metal, and rolling the result into a cylinder were commonly used in the late 19century; their manufacture started in 1876, and they were used from the early 20th century as decoupling capacitors in telecommunications (telephony).

Porcelain was used in the first ceramic capacitors. In the early years of Marconi`s wireless transmitting apparatus porcelain capacitors were used for high voltage and high frequency application in the transmitters. On the receiver side smaller mica capacitors were used for resonant circuits. Mica dielectric capacitors were invented in 1909 by William Dubilier. Prior to World War II, mica was the most common dielectric for capacitors in the United States.

Charles Pollak (born Karol Pollak), the inventor of the first electrolytic capacitors, found out that the oxide layer on an aluminum anode remained stable in a neutral or alkaline electrolyte, even when the power was switched off. In 1896 he was granted U.S. Patent No. 672,913 for an "Electric liquid capacitor with aluminum electrodes." Solid electrolyte tantalum capacitors were invented by Bell Laboratories in the early 1950s as a miniaturized and more reliable low-voltage support capacitor to complement their newly invented transistor.

With the development of plastic materials by organic chemists during the Second World War, the capacitor industry began to replace paper with thinner polymer films. One very early development in film capacitors was described in British Patent 587,953 in 1944.

Last but not least the electric double-layer capacitor (now Supercapacitors) were invented. In 1957 H. Becker developed a "Low voltage electrolytic capacitor with porous carbon electrodes". He believed that the energy was stored as a charge in the carbon pores used in his capacitor as in the pores of the etched foils of electrolytic capacitors. Because the double layer mechanism was not known by him at the time, he wrote in the patent: "It is not known exactly what is taking place in the component if it is used for energy storage, but it leads to an extremely high capacity."

THEORY OF OPERATION

A capacitor consists of two conductors separated by a non-conductive region. The non-conductive region can either be a vacuum or an electrical insulator material known as a dielectric. Examples of dielectric media are glass, air, paper, and even a semiconductor depletion region chemically identical to the conductors. A capacitor is assumed to be self-contained and isolated, with no net electric charge and no influence from any external electric field. The conductors thus hold equal and opposite charges on their facing surfaces, and the dielectric develops an electric field. In SI units, a capacitance of one farad means that one coulomb of charge on each conductor causes a voltage of one volt across the device.

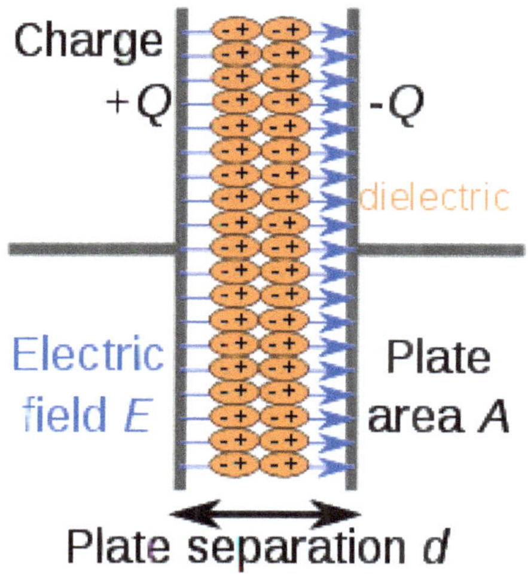

Charge separation in a parallel-plate capacitor causes an internal electric field. A dielectric (orange) reduces the field and increases the capacitance.

A simple demonstration capacitor made of two parallel metal plates, using an air gap as the dielectric.

An ideal capacitor is wholly characterized by a constant capacitance C, defined as the ratio of charge ±Q on each conductor to the voltage V between them:

$$C = \frac{Q}{V}$$

Because the conductors (or plates) are close together, the opposite charges on the conductors attract one another due to their electric fields, allowing the capacitor to store more charge for a given voltage than if the conductors were separated, giving the capacitor a large capacitance.

Sometimes charge build-up affects the capacitor mechanically, causing its capacitance to vary. In this case, capacitance is defined in terms of incremental changes:

$$C = \frac{\mathrm{d}Q}{\mathrm{d}V}$$

Hydraulic Analogy

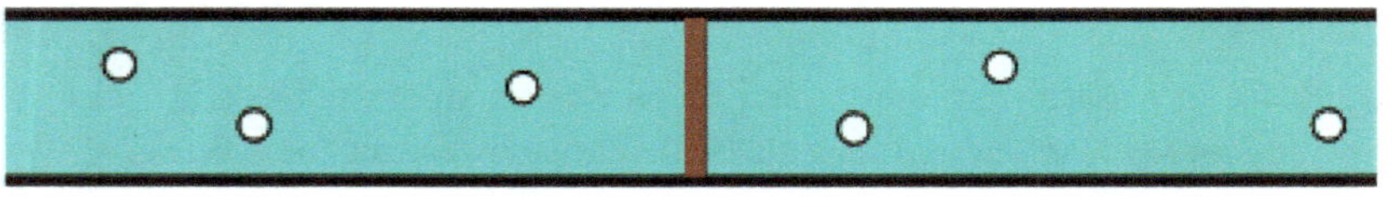

In the hydraulic analogy, a capacitor is analogous to a rubber membrane sealed inside a pipe. This animation illustrates a membrane being repeatedly stretched and un-stretched by the flow of water, which is analogous to a capacitor being repeatedly charged and discharged by the flow of charge.

In the hydraulic analogy, charge carriers flowing through a wire are analogous to water flowing through a pipe. A capacitor is like a rubber membrane sealed inside a pipe. Water molecules cannot pass through the membrane, but some water can move by stretching the membrane. The analogy clarifies a few aspects of capacitors:

- The current alters the charge on a capacitor, just as the flow of water changes the position of the membrane. More specifically, the effect of an electric current is to increase the charge of one plate of the capacitor, and decrease the charge of the other plate by an equal amount. This is just as when water flow moves the rubber membrane, it increases the amount of water on one side of the membrane, and decreases the amount of water on the other side.
- The more a capacitor is charged, the larger its voltage drop; i.e., the more it "pushes back" against the charging current. This is analogous to the fact that the more a membrane is stretched, the more it pushes back on the water.
- Charge can flow "through" a capacitor even though no individual electron can get from one side to the other. This is analogous to water flowing through the pipe even though no water molecule can pass through the rubber membrane. The flow cannot continue in the same direction forever; the capacitor experiences dielectric breakdown, and analogously the membrane will eventually break.

- The capacitance describes how much charge can be stored on one plate of a capacitor for a given "push" (voltage drop). A very stretchy, flexible membrane corresponds to a higher capacitance than a stiff membrane.
- A charged-up capacitor is storing potential energy, analogously to a stretched membrane.

Energy of Electric Field

Work must be done by an external influence to "move" charge between the conductors in a capacitor. When the external influence is removed, the charge separation persists in the electric field and energy is stored to be released when the charge is allowed to return to its equilibrium position. The work done in establishing the electric field, and hence the amount of energy stored, is

$$W = \int_0^Q V(q)\mathrm{d}q = \int_0^Q \frac{q}{C}\mathrm{d}q = \frac{1}{2}\frac{Q^2}{C} = \frac{1}{2}CV^2 = \frac{1}{2}VQ$$

Here Q is the charge stored in the capacitor, V is the voltage across the capacitor, and C is the capacitance.

In the case of a fluctuating voltage V(t), the stored energy also fluctuates and hence power must flow into or out of the capacitor. This power can be found by taking the time derivative of the stored energy:

$$P = \frac{\mathrm{d}W}{dt} = \frac{\mathrm{d}}{dt}\left(\frac{1}{2}CV^2\right) = CV(t)\frac{\mathrm{d}V(t)}{dt}$$

A real capacitor with loss may be modeled as an ideal capacitor that has an Equivalent Series Resistance (ESR) which dissipates power as the capacitor is charged or discharged. For an sinusoidal input voltage the power dissipated due to the ESR is given as:

$$Pd_{rms} = \frac{V_{rms}^2 R_{esr}}{R_{esr}^2 + \left(\frac{1}{2\pi fC}\right)^2}$$

Current–Voltage Relation

The current I(t) through any component in an electric circuit is defined as the rate of flow of a charge Q(t) passing through it, but actual charges—electrons—cannot pass through the dielectric layer of a capacitor. Rather, one electron accumulates on the negative plate for each one that leaves the positive plate, resulting in an electron depletion and consequent positive charge on one electrode that is equal and opposite to the accumulated negative charge on the other. Thus the charge on the electrodes is equal to the integral of the current as well as proportional to the voltage, as discussed above. As with any antiderivative, a constant of integration is added to represent the initial voltage $V(t_c)$. This is the integral form of the capacitor equation:

$$V(t) = \frac{Q(t)}{C} = \frac{1}{C}\int_{t_0}^{t} I(\tau)\mathrm{d}\tau + V(t_0)$$

Taking the derivative of this and multiplying by C yields the derivative form:

$$I(t) = \frac{\mathrm{d}Q(t)}{\mathrm{d}t} = C\frac{\mathrm{d}V(t)}{\mathrm{d}t}$$

The dual of the capacitor is the inductor, which stores energy in a magnetic field rather than an electric field. Its current-voltage relation is obtained by exchanging current and voltage in the capacitor equations and replacing C with the inductance L.

DC Circuits

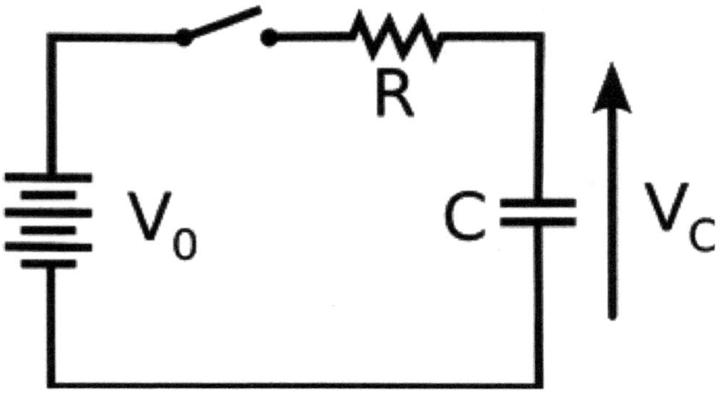

A simple resistor-capacitor circuit demonstrates charging of a capacitor.

A series circuit containing only a resistor, a capacitor, a switch and a constant DC source of voltage V_0 is known as a charging circuit. If the capacitor is initially uncharged while the switch is open, and the switch is closed at t_0, it follows from Kirchhoff's voltage law that

$$V_0 = v_{\text{resistor}}(t) + v_{\text{capacitor}}(t) = i(t)R + \frac{1}{C}\int_{t_0}^{t} i(\tau)\mathrm{d}\tau$$

Taking the derivative and multiplying by C, gives a first-order differential equation:

$$RC\frac{\mathrm{d}i(t)}{\mathrm{d}t} + i(t) = 0$$

At t = 0, the voltage across the capacitor is zero and the voltage across the resistor is V_0. The initial current is then $I(0) = V_0/R$. With this assumption, solving the differential equation yields

$$I(t) = \frac{V_0}{R} e^{-\frac{t}{\tau_0}}$$
$$V(t) = V_0 \left(1 - e^{-\frac{t}{\tau_0}}\right)$$

where τ_0 = RC is the time constant of the system. As the capacitor reaches equilibrium with the source voltage, the voltages across the resistor and the current through the entire circuit decay exponentially. The case of discharging a charged capacitor likewise demonstrates exponential decay, but with the initial capacitor voltage replacing V_0 and the final voltage being zero.

AC Circuits

Impedance, the vector sum of reactance and resistance, describes the phase difference and the ratio of amplitudes between sinusoidally varying voltage and sinusoidally varying current at a given frequency. Fourier analysis allows any signal to be constructed from a spectrum of frequencies, whence the circuit's reaction to the various frequencies may be found. The reactance and impedance of a capacitor are respectively

$$X = -\frac{1}{\omega C} = -\frac{1}{2\pi fC}$$
$$Z = \frac{1}{j\omega C} = -\frac{j}{\omega C} = -\frac{j}{2\pi fC}$$

where j is the imaginary unit and ω is the angular frequency of the sinusoidal signal. The −j phase indicates that the AC voltage V = ZI lags the AC current by 90°: the positive current phase corresponds to increasing voltage as the capacitor charges; zero current corresponds to instantaneous constant voltage, etc.

Impedance decreases with increasing capacitance and increasing frequency. This implies that a higher-frequency signal or a larger capacitor results in a lower voltage amplitude per current amplitude—an AC "short circuit" or AC coupling. Conversely, for very low frequencies, the reactance is high, so that a capacitor is nearly an open circuit in AC analysis—those frequencies have been "filtered out".

Capacitors are different from resistors and inductors in that the impedance is inversely proportional to the defining characteristic; i.e., capacitance.

A capacitor connected to a sinusoidal voltage source causes a displacement current to flow through it. In the case that the voltage source is $V_0\cos(\omega t)$, the displacement current can be expressed as:

$$I = C\frac{dV}{dt} = -\omega C V_0 \sin(\omega t)$$

At sin(ωt) = -1, the capacitor has a maximum (or peak) current whereby $I_0 = \omega C V_0$. The ratio of peak voltage to peak current is due to capacitive reactance (denoted X_C).

$$X_C = \frac{V_0}{I_0} = \frac{V_0}{\omega C V_0} = \frac{1}{\omega C}$$

X_C approaches zero as ω approaches infinity. If X_C approaches 0, the capacitor resembles a short wire that strongly passes current at high frequencies. X_C approaches infinity as ω approaches zero. If X_C approaches infinity, the capacitor resembles an open circuit that poorly passes low frequencies.

The current of the capacitor may be expressed in the form of cosines to better compare with the voltage of the source:

$$I = -I_0 \sin(\omega t) = I_0 \cos(\omega t + 90^\circ)$$

In this situation, the current is out of phase with the voltage by $+\pi/2$ radians or +90 degrees, i.e. the current leads the voltage by 90°.

Laplace circuit Analysis (s-Domain)

When using the Laplace transform in circuit analysis, the impedance of an ideal capacitor with no initial charge is represented in the s domain by:

$$Z(s) = \frac{1}{sC}$$

where

- C is the capacitance, and
- s is the complex frequency.

Parallel-Plate Model

The simplest model capacitor consists of two thin parallel conductive plates separated by a dielectric with permittivity ε . This model may also be used to make qualitative predictions for other device geometries. The plates are considered to extend uniformly over an area A and a charge density $\pm\rho = \pm Q/A$ exists on their surface. Assuming that the length and width of the plates are much greater than their separation d, the electric field near the centre of the device is uniform with the magnitude $E = \rho/\varepsilon$. The voltage is defined as the line integral of the electric field between the plates

$$V = \int_0^d E\,\mathrm{d}z = \int_0^d \frac{\rho}{\varepsilon}\,\mathrm{d}z = \frac{\rho d}{\varepsilon} = \frac{Qd}{\varepsilon A}$$

Solving this for $C = Q/V$ reveals that capacitance increases with area of the plates, and decreases as separation between plates increases.

$$C = \frac{\varepsilon A}{d}$$

The capacitance is therefore greatest in devices made from materials with a high permittivity, large plate area, and small distance between plates.

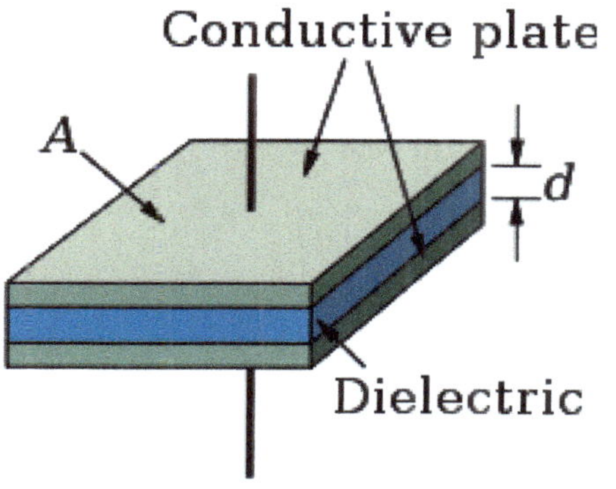

Dielectric is placed between two conducting plates, each of area A and with a separation of d

A parallel plate capacitor can only store a finite amount of energy before dielectric breakdown occurs. The capacitor's dielectric material has a dielectric strength U_d which sets the capacitor's breakdown voltage at $V = V_{bd} = U_d d$. The maximum energy that the capacitor can store is therefore

$$E = \frac{1}{2}CV^2 = \frac{1}{2}\frac{\varepsilon A}{d}(U_d d)^2 = \frac{1}{2}\varepsilon A d U_d^2$$

The maximum energy is a function of dielectric volume, permittivity, and dielectric strength. Changing the plate area and the separation between the plates while maintaining the same volume causes no change of the maximum amount of energy that the capacitor can store, so long as the distance between plates remains much smaller than both the length and width of the plates. In addition, these equations assume that the electric field is entirely concentrated in the dielectric between the plates. In reality there are fringing fields outside the dielectric, for example between the sides of the capacitor plates, which increase the effective capacitance of the capacitor. This is sometimes called parasitic capacitance. For some simple capacitor geometries this additional capacitance term can be calculated analytically. It becomes negligibly small when the ratios of plate width to separation and length to separation are large.

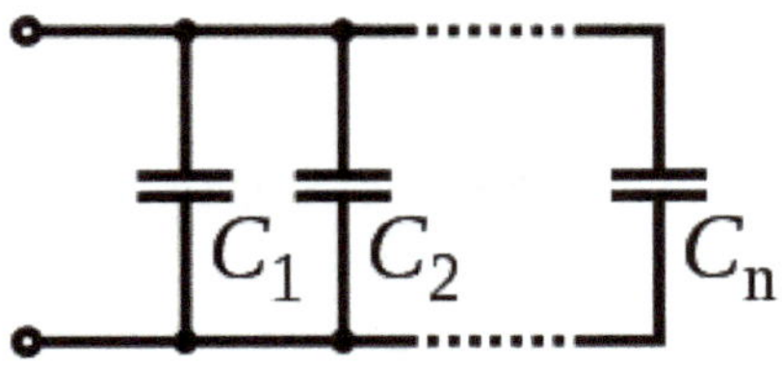

Several capacitors in parallel

Networks

For Capacitors in Parallel

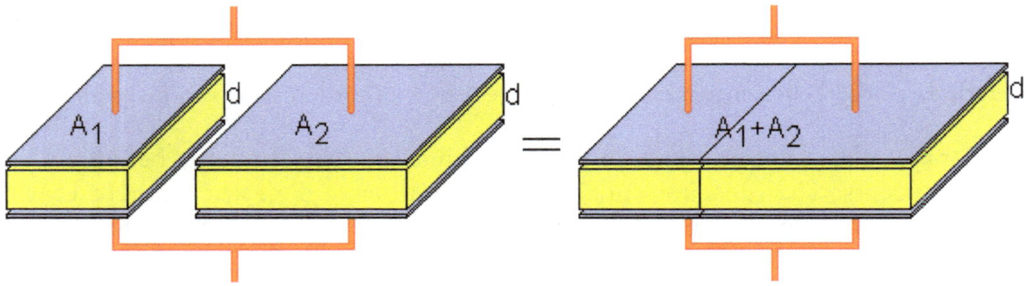

Illustration of the parallel connection of two capacitors.

Capacitors in a parallel configuration each have the same applied voltage. Their capacitances add up. Charge is apportioned among them by size. Using the schematic diagram to visualize parallel plates, it is apparent that each capacitor contributes to the total surface area.

$$C_{eq} = C_1 + C_2 + \cdots + C_n$$

For Capacitors in Series

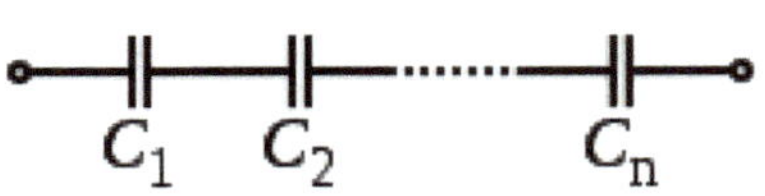

Several capacitors in series

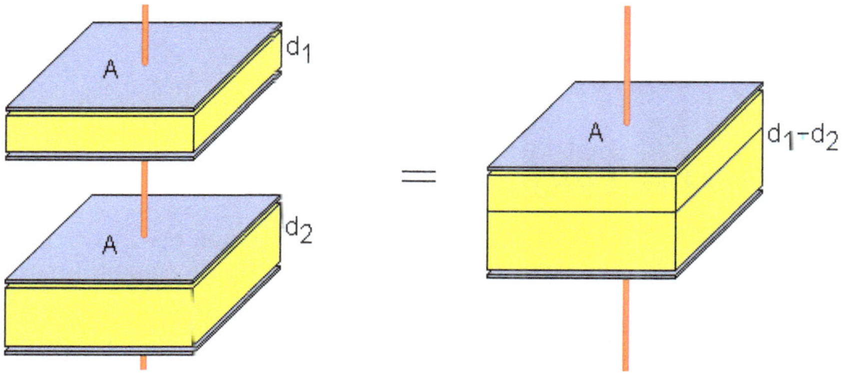

Illustration of the serial connection of two capacitors.

Connected in series, the schematic diagram reveals that the separation distance, not the plate area, adds up. The capacitors each store instantaneous charge build-up equal to that of every other capacitor in the series. The total voltage difference from end to end is apportioned to each capacitor according to the inverse of its capacitance. The entire series acts as a capacitor smaller than any of its components.

$$\frac{1}{C_{\text{eq}}} = \frac{1}{C_1} + \frac{1}{C_2} + \cdots + \frac{1}{C_n}$$

Capacitors are combined in series to achieve a higher working voltage, for example for smoothing a high voltage power supply. The voltage ratings, which are based on plate separation, add up, if capacitance and leakage currents for each capacitor are identical. In such an application, on occasion, series strings are connected in parallel, forming a matrix. The goal is to maximize the energy storage of the network without overloading any capacitor. For high-energy storage with capacitors in series, some safety considerations must be applied to ensure one capacitor failing and leaking current does not apply too much voltage to the other series capacitors.

Series connection is also sometimes used to adapt polarized electrolytic capacitors for bipolar AC use.

Voltage Distribution in Parallel-to-Series Networks

To model the distribution of voltages from a single charged capacitor (A) connected in parallel to a chain of capacitors in series (B_n):

$$(volts)A_{\text{eq}} = A\left(1 - \frac{1}{n+1}\right)$$

$$(volts)B_{1..n} = \frac{A}{n}\left(1 - \frac{1}{n+1}\right)$$

$$A - B = 0$$

Note: This is only correct if all capacitance values are equal.

The power transferred in this arrangement is:

$$P = \frac{1}{R} \cdot \frac{1}{n+1} A_{\text{volts}} \left(A_{\text{farads}} + B_{\text{farads}}\right)$$

NON-IDEAL BEHAVIOR

Capacitors deviate from the ideal capacitor equation in a number of ways. Some of these, such as leakage current and parasitic effects are linear, or can be analyzed as nearly linear, and can be dealt with by adding virtual components to the equivalent circuit of an ideal capacitor. The usual methods of network analysis can then be applied. In other cases, such as with breakdown voltage, the effect is non-linear and ordinary (normal, e.g., linear) network analysis cannot be used, the effect must be dealt with separately. There is yet another group, which may be linear but invalidate the assumption in the analysis that capacitance is a constant. Such an example is temperature dependence. Finally, combined parasitic effects such as inherent inductance, resistance, or dielectric losses can exhibit non-uniform behavior at variable frequencies of operation.

Breakdown Voltage

Above a particular electric field, known as the dielectric strength E_{ds}, the dielectric in a capacitor becomes conductive. The voltage at which this occurs is called the breakdown voltage of the device, and is given by the product of the dielectric strength and the separation between the conductors,

$$V_{\text{bd}} = E_{\text{ds}} d$$

The maximum energy that can be stored safely in a capacitor is limited by the breakdown voltage. Due to the scaling of capacitance and breakdown voltage with dielectric thickness, all capacitors made with a particular dielectric have approximately equal maximum energy density, to the extent that the dielectric dominates their volume.

For air dielectric capacitors the breakdown field strength is of the order 2 to 5 MV/m; for mica the breakdown is 100 to 300 MV/m; for oil, 15 to 25 MV/m; it can be much less when other materials are used for the dielectric. The dielectric is used in very thin layers and so absolute breakdown voltage of capacitors is limited. Typical ratings for capacitors used for general electronics applications range from a few volts to 1 kV. As the voltage increases, the dielectric must be thicker, making high-voltage capacitors larger per capacitance than those rated for lower voltages. The breakdown voltage is critically affected by factors such as the geometry of the capacitor conductive parts; sharp edges or points increase the electric field strength at that point and can lead to a local breakdown. Once this starts to happen, the breakdown quickly tracks through the dielectric until it reaches the opposite plate, leaving carbon behind and causing a short (or relatively low resistance) circuit. The results can be explosive as the short in the capacitor draws current from the surrounding circuitry and dissipates the energy.

The usual breakdown route is that the field strength becomes large enough to pull electrons in the dielectric from their atoms thus causing conduction. Other scenarios are possible, such as impurities in the dielectric, and, if the dielectric is of a crystalline nature, imperfections in the crystal structure can result in an avalanche breakdown as seen in semi-conductor devices. Breakdown voltage is also affected by pressure, humidity and temperature.

Equivalent Circuit

An ideal capacitor only stores and releases electrical energy, without dissipating any. In reality, all capacitors have imperfections within the capacitor's material that create resistance. This is specified as the equivalent series resistance or **ESR** of a component. This adds a real component to the impedance:

$$Z_C = Z + R_{ESR} = \frac{1}{j\omega C} + R_{ESR}$$

As frequency approaches infinity, the capacitive impedance (or reactance) approaches zero and the ESR becomes significant. As the reactance becomes negligible, power dissipation approaches $P_{RMS} = V_{RMS}^2 / R_{ESR}$.

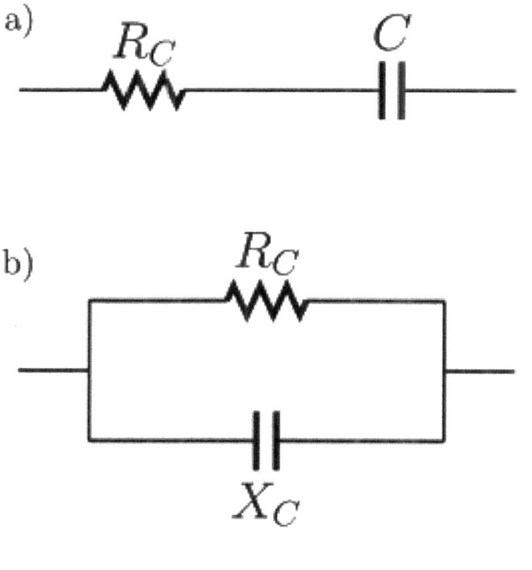

Two different circuit models of a real capacitor

Similarly to ESR, the capacitor's leads add equivalent series inductance or **ESL** to the component. This is usually significant only at relatively high frequencies. As inductive reactance is positive and increases with frequency, above a certain frequency capacitance is canceled by inductance. High-frequency engineering involves accounting for the inductance of all connections and components.

If the conductors are separated by a material with a small conductivity rather than a perfect dielectric, then a small leakage current flows directly between them. The capacitor therefore has a finite parallel resistance, and slowly discharges over time (time may vary greatly depending on the capacitor material and quality).

Q Factor

The quality factor (or Q) of a capacitor is the ratio of its reactance to its resistance at a given frequency, and is a measure of its efficiency. The higher the Q factor of the capacitor, the closer it approaches the behavior of an ideal, lossless, capacitor.

The Q factor of a capacitor can be found through the following formula:

$$Q = \frac{X_C}{R_C} = \frac{1}{\omega C R_C},$$

where ω is angular frequency, C is the capacitance, X_C is the capacitive reactance, and R_C is the series resistance of the capacitor.

Ripple Current

Ripple current is the AC component of an applied source (often a switched-mode power supply) whose frequency may be constant or varying. Ripple current causes heat to be generated within the capacitor due to the dielectric losses caused by the changing field strength together with the current flow across the slightly resistive supply lines or the electrolyte in the capacitor. The equivalent series resistance (ESR) is the amount of internal series resistance one would add to a perfect capacitor to model this.

Some types of capacitors, primarily tantalum and aluminum electrolytic capacitors, as well as some film capacitors have a specified rating value for maximum ripple current.

- Tantalum electrolytic capacitors with solid manganese dioxide electrolyte are limited by ripple current and generally have the highest ESR ratings in the capacitor family. Exceeding their ripple limits can lead to shorts and burning parts.
- Aluminum electrolytic capacitors, the most common type of electrolytic, suffer a shortening of life expectancy at higher ripple currents. If ripple current exceeds the rated value of the capacitor, it tends to result in explosive failure.
- Ceramic capacitors generally have no ripple current limitation and have some of the lowest ESR ratings.
- Film capacitors have very low ESR ratings but exceeding rated ripple current may cause degradation failures.

Capacitance Instability

The capacitance of certain capacitors decreases as the component ages. In ceramic capacitors, this is caused by degradation of the dielectric. The type of dielectric, ambient operating and storage temperatures are the most significant aging factors, while the operating voltage has a smaller effect. The aging process may be reversed by heating the component above the Curie point. Aging is fastest near the beginning of life of the component, and the device stabilizes over time. Electrolytic capacitors age as the electrolyte evaporates. In contrast with ceramic capacitors, this occurs towards the end of life of the component.

Temperature dependence of capacitance is usually expressed in parts per million (ppm) per °C. It can usually be taken as a broadly linear function but can be noticeably non-linear at the temperature extremes. The temperature coefficient can be either positive or negative, sometimes even amongst different samples of the same type. In other words, the spread in the range of temperature coefficients can encompass zero.

Capacitors, especially ceramic capacitors, and older designs such as paper capacitors, can absorb sound waves resulting in a microphonic effect. Vibration moves the plates, causing the capacitance to vary, in turn inducing AC current. Some dielectrics also generate piezoelectricity. The resulting interference is especially problematic in audio applications, potentially causing feedback or unintended recording. In the reverse microphonic effect, the varying electric field between the capacitor plates exerts a physical force, moving them as a speaker. This can generate audible sound, but drains energy and stresses the dielectric and the electrolyte, if any.

Current and Voltage Reversal

Current reversal occurs when the current changes direction. Voltage reversal is the change of polarity in a circuit. Reversal is generally described as the percentage of the maximum rated voltage that reverses polarity. In DC circuits, this is usually less than 100%, often in the range of 0 to 90%, whereas AC circuits experience 100% reversal.

In DC circuits and pulsed circuits, current and voltage reversal are affected by the damping of the system. Voltage reversal is encountered in RLC circuits that are under-damped. The current and voltage reverse direction, forming a harmonic oscillator between the inductance and capacitance. The current and voltage tends to oscillate and may reverse direction several times, with each peak being lower than the previous, until the system reaches an equilibrium. This is often referred to as ringing. In comparison, critically damped or over-damped systems usually do not experience a voltage reversal. Reversal is also encountered in AC circuits, where the peak current is equal in each direction.

For maximum life, capacitors usually need to be able to handle the maximum amount of reversal that a system may experience. An AC circuit experiences 100% voltage reversal, while under-damped DC circuits experience less than 100%. Reversal creates excess electric fields in the dielectric, causes excess heating of both the dielectric and the conductors, and can dramatically shorten the life expectancy of the capacitor. Reversal ratings often affect the design considerations for the capacitor, from the choice of dielectric materials and voltage ratings to the types of internal connections used.

Dielectric Absorption

Capacitors made with any type of dielectric material show some level of "dielectric absorption" or "soakage". On discharging a capacitor and disconnecting it, after a short time it may develop a voltage due to hysteresis in the dielectric. This effect is objectionable in applications such as precision sample and hold circuits or timing circuits. The level of absorption depends on many factors, from design considerations to charging time, since the absorption is a time-dependent process. However, the primary factor is the type of dielectric material. Capacitors such as tantalum electrolytic or polysulfone film exhibit relatively high absorption, while polystyrene or Teflon allow very small levels of absorption. In some capacitors where dangerous voltages and energies exist, such as in flashtubes, television sets, and defibrillators, the dielectric absorption can recharge the capacitor to hazardous voltages after it has been shorted or discharged. Any capacitor containing over 10 joules of energy is generally considered hazardous, while 50 joules or higher is potentially lethal. A capacitor may regain anywhere from 0.01 to 20% of its original charge over a period of several minutes, allowing a seemingly safe capacitor to become surprisingly dangerous.

Leakage

Leakage is equivalent to a resistor in parallel with the capacitor. Constant exposure to heat can cause dielectric breakdown and excessive leakage, a problem often seen in older vacuum tube circuits, particularly where oiled paper and foil capacitors were used. In many vacuum tube circuits, interstage coupling capacitors are used to conduct a varying signal from the plate of one tube to the grid circuit of the next stage. A leaky capacitor can cause the grid circuit voltage to be raised from its normal bias setting, causing excessive current or signal distortion in the downstream tube. In power amplifiers this can cause the plates to glow red, or current limiting resistors to overheat, even fail. Similar considerations apply to component fabricated solid-state (transistor) amplifiers, but owing to lower heat production and the use of modern polyester dielectric barriers this once-common problem has become relatively rare.

Electrolytic Failure from Disuse

Aluminum electrolytic capacitors are conditioned when manufactured by applying a voltage sufficient to initiate the proper internal chemical state. This state is maintained by regular use of the equipment. If a system using electrolytic capacitors is unused for a long period of time it can lose its conditioning. Sometimes they fail with a short circuit when next operated.

Capacitor Types

Practical capacitors are available commercially in many different forms. The type of internal dielectric, the structure of the plates and the device packaging all strongly affect the characteristics of the capacitor, and its applications.

Values available range from very low (picofarad range; while arbitrarily low values are in principle possible, stray (parasitic) capacitance in any circuit is the limiting factor) to about 5 kF supercapacitors.

Above approximately 1 microfarad electrolytic capacitors are usually used because of their small size and low cost compared with other types, unless their relatively poor stability, life and polarized nature make them unsuitable. Very high capacity supercapacitors use a porous carbon-based electrode material.

Dielectric Materials

Most capacitors have a dielectric spacer, which increases their capacitance compared to air or a vacuum. In order to maximize the charge that a capacitor can hold, the dielectric material needs to have as high a permittivity as possible, while also having as high a breakdown voltage as possible. The dielectric also needs to have as low a loss with frequency as possible.

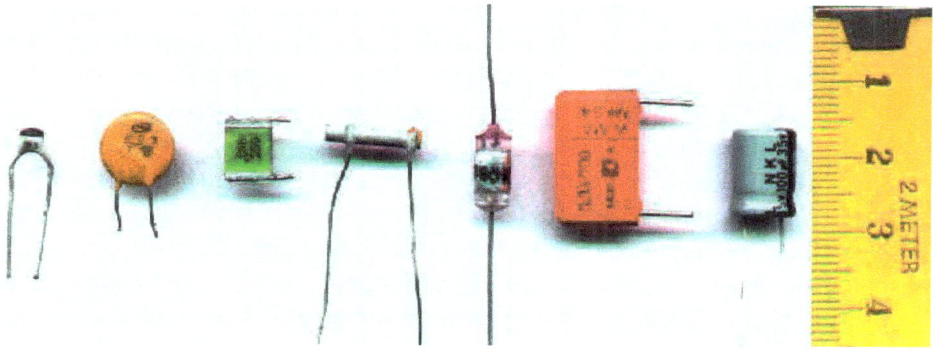

Capacitor materials. From left: multilayer ceramic, ceramic disc, multilayer polyester film, tubular ceramic, polystyrene, metalized polyester film, aluminum electrolytic. Major scale divisions are in centimetres.

However, low value capacitors are available with a vacuum between their plates to allow extremely high voltage operation and low losses. Variable capacitors with their plates open to the atmosphere were commonly used in radio tuning circuits. Later designs use polymer foil dielectric between the moving and stationary plates, with no significant air space between the plates.

Several solid dielectrics are available, including paper, plastic, glass, mica and ceramic.

Paper was used extensively in older capacitors and offers relatively high voltage performance. However, paper absorbs moisture, and has been largely replaced by plastic film capacitors.

Most of the plastic films now used offer better stability and ageing performance than such older dielectrics such as oiled paper, which makes them useful in timer circuits, although they may be limited to relatively low operating temperatures and frequencies, because of the limitations of the plastic film being used. Large plastic film capacitors are used extensively in suppression circuits, motor start circuits, and power factor correction circuits.

Ceramic capacitors are generally small, cheap and useful for high frequency applications, although their capacitance varies strongly with voltage and temperature and they age poorly. They can also suffer from the piezoelectric effect. Ceramic capacitors are broadly categorized as class 1 dielectrics, which have predictable variation of capacitance with temperature or class 2 dielectrics, which can operate at higher voltage. Modern multilayer ceramics are usually quite small, but some types have inherently wide value tolerances, microphonic issues, and are usually physically brittle.

Glass and mica capacitors are extremely reliable, stable and tolerant to high temperatures and voltages, but are too expensive for most mainstream applications.

Electrolytic capacitors and supercapacitors are used to store small and larger amounts of energy, respectively, ceramic capacitors are often used in resonators, and parasitic capacitance occurs in circuits wherever the simple conductor-insulator-conductor structure is formed unintentionally by the configuration of the circuit layout.

Electrolytic capacitors use an aluminum or tantalum plate with an oxide dielectric layer. The second electrode is a liquid electrolyte, connected to the circuit by another foil plate. Electrolytic capacitors offer very high capacitance but suffer from poor tolerances, high instability, gradual loss of capacitance especially when subjected to heat, and high leakage current. Poor quality capacitors may leak electrolyte, which is harmful to printed circuit boards. The conductivity of the electrolyte drops at low temperatures, which increases equivalent series resistance. While widely used for power-supply conditioning, poor high-frequency characteristics make them unsuitable for many applications. Electrolytic capacitors suffer from self-degradation if unused for a period (around a year), and when full power is applied may short circuit, permanently damaging the capacitor and usually blowing a fuse or causing failure of rectifier diodes. For example, in older equipment, this may cause arcing in rectifier tubes. They can be restored before use by gradually applying the operating voltage, often performed on antique vacuum tube equipment over a period of thirty minutes by using a variable transformer to supply AC power. The use of this technique may be less satisfactory for some solid state equipment, which may be damaged by operation below its normal power range, requiring that the power supply first be isolated from the consuming circuits. Such remedies may not be applicable to modern high-frequency power supplies as these produce full output voltage even with reduced input.

Tantalum capacitors offer better frequency and temperature characteristics than aluminum, but higher dielectric absorption and leakage.

Polymer capacitors (OS-CON, OC-CON, KO, AO) use solid conductive polymer (or polymerized organic semiconductor) as electrolyte and offer longer life and lower ESR at higher cost than standard electrolytic capacitors.

A feedthrough capacitor is a component that, while not serving as its main use, has capacitance and is used to conduct signals through a conductive sheet.

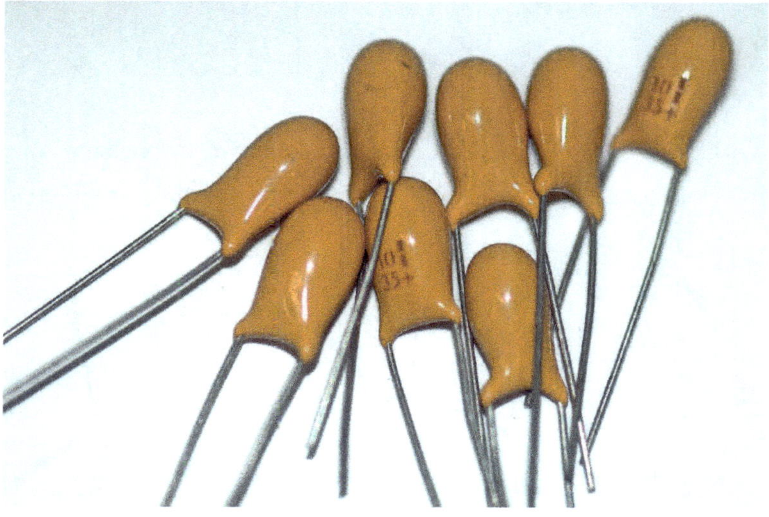

Solid electrolyte, resin-dipped 10 μF 35 V tantalum capacitors. The + sign indicates the positive lead.

Several other types of capacitor are available for specialist applications. Supercapacitors store large amounts of energy. Supercapacitors made from carbon aerogel, carbon nanotubes, or highly porous electrode materials, offer extremely high capacitance (up to 5 kF as of 2010) and can be used in some applications instead of rechargeable batteries. Alternating current capacitors are specifically designed to work on line (mains) voltage AC power circuits. They are commonly used in electric motor circuits and are often designed to handle large currents, so they tend to be physically large. They are usually ruggedly packaged, often in metal cases that can be easily grounded/earthed. They also are designed with direct current breakdown voltages of at least five times the maximum AC voltage.

Voltage-Dependent Capacitors

The dielectric constant for a number of very useful dielectrics changes as a function of the applied electrical field, for example ferroelectric materials, so the capacitance for these devices is more complex. For example, in charging such a capacitor the differential increase in voltage with charge is governed by:

$$dQ = C(V)\,dV$$

where the voltage dependence of capacitance, C(V), suggests that the capacitance is a function of the electric field strength, which in a large area parallel plate device is given by $\varepsilon = V/d$. This field polarizes the dielectric, which polarization, in the case of a ferroelectric, is a nonlinear S-shaped function of the electric field, which, in the case of a large area parallel plate device, translates into a capacitance that is a nonlinear function of the voltage.

Corresponding to the voltage-dependent capacitance, to charge the capacitor to voltage V an integral relation is found:

$$Q = \int_0^V C(V)\,dV$$

which agrees with Q = CV only when C does not depend on voltage V.

By the same token, the energy stored in the capacitor now is given by

$$dW = Q\,dV = \left[\int_0^V dV'\,C(V')\right] dV\,.$$

Integrating:

$$W = \int_0^V dV \int_0^V dV'\,C(V') = \int_0^V dV' \int_{V'}^V dV\,C(V') = \int_0^V dV'\,(V - V')\,C(V')\,,$$

where interchange of the order of integration is used.

The nonlinear capacitance of a microscope probe scanned along a ferroelectric surface is used to study the domain structure of ferroelectric materials.

Another example of voltage dependent capacitance occurs in semiconductor devices such as semiconductor diodes, where the voltage dependence stems not from a change in dielectric constant but in a voltage dependence of the spacing between the charges on the two sides of the capacitor. This effect is intentionally exploited in diode-like devices known as varicaps.

Frequency-Dependent Capacitors

If a capacitor is driven with a time-varying voltage that changes rapidly enough, at some frequency the polarization of the dielectric cannot follow the voltage. As an example of the origin of this mechanism, the internal microscopic dipoles contributing to the dielectric constant cannot move instantly, and so as frequency of an applied alternating voltage increases, the dipole response is limited and the dielectric constant diminishes. A changing dielectric constant with frequency is referred to as dielectric dispersion, and is governed by dielectric relaxation processes, such as Debye relaxation. Under transient conditions, the displacement field can be expressed as:

$$\boldsymbol{D}(t) = \varepsilon_0 \int_{-\infty}^{t} \varepsilon_r(t - t')\boldsymbol{E}(t')\,dt',$$

indicating the lag in response by the time dependence of ε_r, calculated in principle from an underlying microscopic analysis, for example, of the dipole behavior in the dielectric. See, for example, linear response function. The integral extends over the entire past history up to the present time. A Fourier transform in time then results in:

$$\boldsymbol{D}(\omega) = \varepsilon_0 \varepsilon_r(\omega)\boldsymbol{E}(\omega)\,,$$

where $\varepsilon_r(\omega)$ is now a complex function, with an imaginary part related to absorption of energy from the field by the medium. The capacitance, being proportional to the dielectric constant, also exhibits this frequency behavior. Fourier transforming Gauss›s law with this form for displacement field:

$$I(\omega) = j\omega Q(\omega) = j\omega \oint_{\Sigma} \boldsymbol{D}(\boldsymbol{r},\ \omega) \cdot d\boldsymbol{\Sigma}$$

$$= [G(\omega) + j\omega C(\omega)]\, V(\omega) = \frac{V(\omega)}{Z(\omega)},$$

where j is the imaginary unit, $V(\omega)$ is the voltage component at angular frequency ω, $G(\omega)$ is the real part of the current, called the conductance, and $C(\omega)$ determines the imaginary part of the current and is the capacitance. $Z(\omega)$ is the complex impedance.

When a parallel-plate capacitor is filled with a dielectric, the measurement of dielectric properties of the medium is based upon the relation:

$$\varepsilon_r(\omega) = \varepsilon_r'(\omega) - j\varepsilon_r''(\omega) = \frac{1}{j\omega Z(\omega) C_0} = \frac{C_{\mathrm{cmplx}}(\omega)}{C_0},$$

where a single prime denotes the real part and a double prime the imaginary part, $Z(\omega)$ is the complex impedance with the dielectric present, $C_{\mathrm{cmplx}}(\omega)$ is the so-called complex capacitance with the dielectric present, and C_0 is the capacitance without the dielectric. (Measurement "without the dielectric" in principle means measurement in free space, an unattainable goal inasmuch as even the quantum vacuum is predicted to exhibit nonideal behavior, such as dichroism. For practical purposes, when measurement errors are taken into account, often a measurement in terrestrial vacuum, or simply a calculation of C_0, is sufficiently accurate.)

Using this measurement method, the dielectric constant may exhibit a resonance at certain frequencies corresponding to characteristic response frequencies (excitation energies) of contributors to the dielectric constant. These resonances are the basis for a number of experimental techniques for detecting defects. The conductance method measures absorption as a function of frequency. Alternatively, the time response of the capacitance can be used directly, as in deep-level transient spectroscopy.

Another example of frequency dependent capacitance occurs with MOS capacitors, where the slow generation of minority carriers means that at high frequencies the capacitance measures only the majority carrier response, while at low frequencies both types of carrier respond.

At optical frequencies, in semiconductors the dielectric constant exhibits structure related to the band structure of the solid. Sophisticated modulation spectroscopy measurement methods based upon modulating the crystal structure by pressure or by other stresses and observing the related changes in absorption or reflection of light have advanced our knowledge of these materials.

Structure

The arrangement of plates and dielectric has many variations depending on the desired ratings of the capacitor. For small values of capacitance (microfarads and less), ceramic disks use metallic coatings, with wire leads bonded to the coating. Larger values can be made by multiple stacks of plates and disks. Larger value capacitors usually use a metal foil or metal film layer deposited on the surface of a dielectric film to make the plates, and a dielectric film of impregnated paper or plastic – these are rolled up to save space. To reduce the series resistance and inductance for long plates, the plates and dielectric are staggered so that connection is made at the common edge of the rolled-up plates, not at the ends of the foil or metalized film strips that comprise the plates.

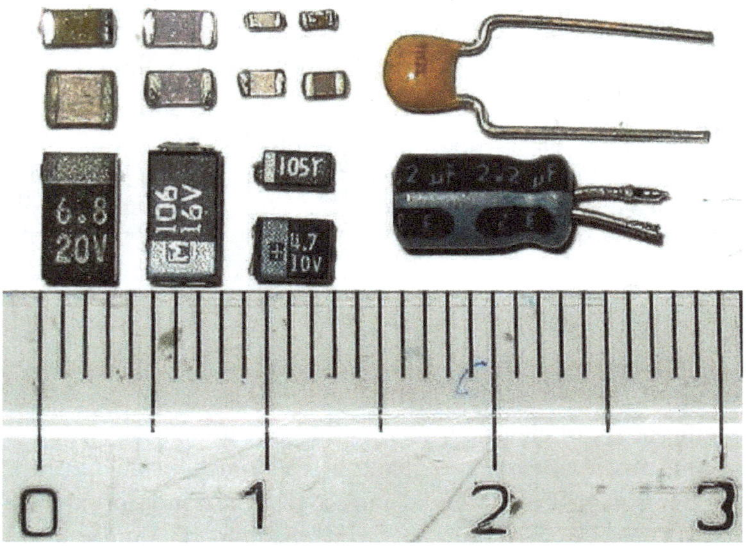

Capacitor packages: SMD ceramic at top left; SMD tantalum at bottom left; through-hole tantalum at top right; through-hole electrolytic at bottom right. Major scale divisions are cm.

The assembly is encased to prevent moisture entering the dielectric – early radio equipment used a cardboard tube sealed with wax. Modern paper or film dielectric capacitors are dipped in a hard thermoplastic. Large capacitors for high-voltage use may have the roll form compressed to fit into a rectangular metal case, with bolted terminals and bushings for connections. The dielectric in larger capacitors is often impregnated with a liquid to improve its properties.

Capacitors may have their connecting leads arranged in many configurations, for example axially or radially. "Axial" means that the leads are on a common axis, typically the axis of the capacitor's cylindrical body – the leads extend from opposite ends. Radial leads might more accurately be referred to as tandem; they are rarely actually aligned along radii of the body's circle, so the term is inexact, although universal. The leads (until bent) are usually in planes parallel to that of the flat body of the capacitor, and extend in the same direction; they are often parallel as manufactured.

Several axial-lead electrolytic capacitors

Small, cheap discoidal ceramic capacitors have existed since the 1930s, and remain in widespread use. Since the 1980s, surface mount packages for capacitors have been widely used. These packages are extremely small and lack connecting leads, allowing them to be soldered directly onto the surface of printed circuit boards. Surface mount components avoid undesirable high-frequency effects due to the leads and simplify automated assembly, although manual handling is made difficult due to their small size.

Mechanically controlled variable capacitors allow the plate spacing to be adjusted, for example by rotating or sliding a set of movable plates into alignment with a set of stationary plates. Low cost variable capacitors squeeze together alternating layers of aluminum and plastic with a screw. Electrical control of capacitance is achievable with varactors (or varicaps), which are reverse-biased semiconductor diodes whose depletion region width varies with applied voltage. They are used in phase-locked loops, amongst other applications.

CAPACITOR MARKINGS

Most capacitors have numbers printed on their bodies to indicate their electrical characteristics. Larger capacitors like electrolytics usually display the actual capacitance together with the unit, for example, 220 µF. Smaller capacitors like ceramics, however, use a shorthand-notation consisting of three digits and a letter, where the digits indicate the capacitance in pF, calculated as $XY \times 10^Z$ for digits XYZ, and the letter indicates the tolerance. Common tolerance indications are J, K, and M for ±5%, ±10%, and ±20%, respectively.

Additionally, the capacitor may be labeled with its working voltage, temperature and other relevant characteristics.

For typographical reasons, some manufacturers print MF on capacitors to indicate microfarads (µF).

Example

A capacitor labeled or designated as 473K 330V has a capacitance of 47×10^3 pF = 47 nF (±10%) with a maximum working voltage of 330 V. The working voltage of a capacitor is nominally the highest voltage that may be applied across it without undue risk of breaking down the dielectric layer.

Letter and Digit Code

The notation to state a capacitor's value in a circuit diagram varies. The letter and digit code for capacitance values following IEC 60062 and BS 1852 avoids using a decimal separator and replaces the decimal separator with the SI prefix symbol for the particular value (and the letter F for weight 1). Example: 4n7 for 4.7 nF or 2F2 for 2.2 F.

Historical

In the past, alternate capacitance subunits were used in historical electronic books; "mfd" and "mf" for microfarad (μF); "mmfd", "mmf", "μμF" for picofarad (pF); but are rarely used any more.

APPLICATIONS

This mylar-film, oil-filled capacitor has very low inductance and low resistance, to provide the high-power (70 megawatt) and high speed (1.2 microsecond) discharge needed to operate a dye laser.

Energy Storage

A capacitor can store electric energy when disconnected from its charging circuit, so it can be used like a temporary battery, or like other types of rechargeable energy storage system. Capacitors are commonly used in electronic devices to maintain power supply while batteries are being changed. (This prevents loss of information in volatile memory.)

Conventional capacitors provide less than 360 joules per kilogram of specific energy, whereas a conventional alkaline battery has a density of 590 kJ/kg. There is an intermediate solution: Supercapacitors, which can accept and deliver charge much faster than batteries, and tolerate many more charge and discharge cycles than rechargeable batteries. They are however 10 times larger than conventional batteries for a given charge.

In car audio systems, large capacitors store energy for the amplifier to use on demand. Also for a flash tube a capacitor is used to hold the high voltage.

Digital Memory

In the 1930s, John Atanasoff applied the principle of energy storage in capacitors to construct dynamic digital memories for the first binary computers that used electron tubes for logic.

Pulsed Power and Weapons

Groups of large, specially constructed, low-inductance high-voltage capacitors (capacitor banks) are used to supply huge pulses of current for many pulsed power applications. These include electromagnetic forming, Marx generators, pulsed lasers (especially TEA lasers), pulse forming networks, radar, fusion research, and particle accelerators.

Large capacitor banks (reservoir) are used as energy sources for the exploding-bridgewire detonators or slapper detonators in nuclear weapons and other specialty weapons. Experimental work is under way using banks of capacitors as power sources for electromagnetic armour and electromagnetic railguns and coilguns.

Power Conditioning

Reservoir capacitors are used in power supplies where they smooth the output of a full or half wave rectifier. They can also be used in charge pump circuits as the energy storage element in the generation of higher voltages than the input voltage.

A 10,000 microfarad capacitor in an amplifier power supply

Capacitors are connected in parallel with the power circuits of most electronic devices and larger systems (such as factories) to shunt away and conceal current fluctuations from the primary power source to provide a "clean" power supply for signal or control circuits. Audio equipment, for example, uses several capacitors in this way, to shunt away power line hum before it gets into the signal circuitry. The capacitors act as a local reserve for the DC power source, and bypass AC currents from the power supply. This is used in car audio applications, when a stiffening capacitor compensates for the inductance and resistance of the leads to the lead-acid car battery.

Power Factor Correction

A high-voltage capacitor bank used for power factor correction on a power transmission system

In electric power distribution, capacitors are used for power factor correction. Such capacitors often come as three capacitors connected as a three phase load. Usually, the values of these capacitors are given not in farads but rather as a reactive power in volt-amperes reactive (var). The purpose is to counteract inductive loading from devices like electric motors and transmission lines to make the load appear to be mostly resistive. Individual motor or lamp loads may have capacitors for power factor correction, or larger sets of capacitors (usually with automatic switching devices) may be installed at a load center within a building or in a large utility substation.

Suppression and Coupling

Signal Coupling

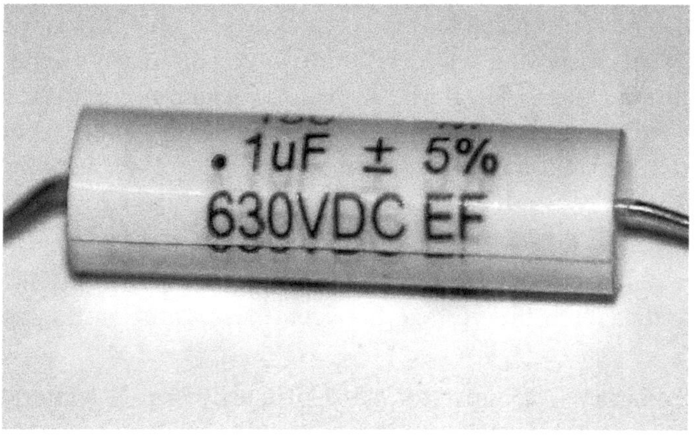

Polyester film capacitors are frequently used as coupling capacitors.

Because capacitors pass AC but block DC signals (when charged up to the applied dc voltage), they are often used to separate the AC and DC components of a signal. This method is known as AC coupling or "capacitive coupling". Here, a large value of capacitance, whose value need not be accurately controlled, but whose reactance is small at the signal frequency, is employed.

Decoupling

A decoupling capacitor is a capacitor used to protect one part of a circuit from the effect of another, for instance to suppress noise or transients. Noise caused by other circuit elements is shunted through the capacitor, reducing the effect they have on the rest of the circuit. It is most commonly used between the power supply and ground. An alternative name is bypass capacitor as it is used to bypass the power supply or other high impedance component of a circuit.

Decoupling capacitors need not always be discrete components. Capacitors used in these applications may be built into a printed circuit board, between the various layers. These are often referred to as embedded capacitors. The layers in the board contributing to the capacitive properties also function as power and ground planes, and have a dielectric in between them, enabling them to operate as a parallel plate capacitor.

High-Pass and Low-Pass Filters

Noise Suppression, Spikes, and Snubbers

When an inductive circuit is opened, the current through the inductance collapses quickly, creating a large voltage across the open circuit of the switch or relay. If the inductance is large enough, the energy may generate a spark, causing the contact points to oxidize, deteriorate, or sometimes weld together, or destroying a solid-state switch. A snubber capacitor across the newly opened circuit creates a path for this impulse to bypass the contact points, thereby preserving their life; these were commonly found in contact breaker ignition systems, for instance. Similarly, in smaller scale circuits, the spark may not be enough to damage the switch but may still radiate undesirable radio frequency interference (RFI), which a filter capacitor absorbs. Snubber capacitors are usually employed with a low-value resistor in series, to dissipate energy and minimize RFI. Such resistor-capacitor combinations are available in a single package.

Capacitors are also used in parallel to interrupt units of a high-voltage circuit breaker in order to equally distribute the voltage between these units. In this case they are called grading capacitors.

In schematic diagrams, a capacitor used primarily for DC charge storage is often drawn vertically in circuit diagrams with the lower, more negative, plate drawn as an arc. The straight plate indicates the positive terminal of the device, if it is polarized.

Motor Starters

In single phase squirrel cage motors, the primary winding within the motor housing is not capable of starting a rotational motion on the rotor, but is capable of sustaining one. To start the motor, a secondary "start" winding has a series non-polarized starting capacitor to introduce a lead in the sinusoidal current. When the secondary (start) winding is placed at an angle with respect to the primary (run) winding, a rotating electric field is created. The force of the rotational field is not constant, but is sufficient to start the rotor spinning. When the rotor comes close to operating speed, a centrifugal switch (or current-sensitive relay in series with the main winding) disconnects the capacitor. The start capacitor is typically mounted to the side of the motor housing. These are called capacitor-start motors, that have relatively high starting torque. Typically they can have up-to four times as much starting torque than a split-phase motor and are used on applications such as compressors, pressure washers and any small device requiring high starting torques.

Capacitor-run induction motors have a permanently connected phase-shifting capacitor in series with a second winding. The motor is much like a two-phase induction motor.

Motor-starting capacitors are typically non-polarized electrolytic types, while running capacitors are conventional paper or plastic film dielectric types.

Signal Processing

The energy stored in a capacitor can be used to represent information, either in binary form, as in DRAMs, or in analogue form, as in analog sampled filters and CCDs. Capacitors can be used in analog circuits as components of integrators or more complex filters and in negative feedback loop stabilization. Signal processing circuits also use capacitors to integrate a current signal.

Tuned Circuits

Capacitors and inductors are applied together in tuned circuits to select information in particular frequency bands. For example, radio receivers rely on variable capacitors to tune the station frequency. Speakers use passive analog crossovers, and analog equalizers use capacitors to select different audio bands.

The resonant frequency f of a tuned circuit is a function of the inductance (L) and capacitance (C) in series, and is given by:

$$f = \frac{1}{2\pi\sqrt{LC}}$$

where L is in henries and C is in farads.

Sensing

Most capacitors are designed to maintain a fixed physical structure. However, various factors can change the structure of the capacitor, and the resulting change in capacitance can be used to sense those factors.

Changing the dielectric:

The effects of varying the characteristics of the dielectric can be used for sensing purposes. Capacitors with an exposed and porous dielectric can be used to measure humidity in air. Capacitors are used to accurately measure the fuel level in airplanes; as the fuel covers more of a pair of plates, the circuit capacitance increases. Squeezing the dielectric can change a capacitor at a few tens of bar pressure sufficiently that it can be used as a pressure sensor. A selected, but otherwise standard, polymer dielectric capacitor, when immersed in a compatible gas or liquid, can work usefully as a very low cost pressure sensor up to many hundreds of bar.

Changing the distance between the plates:

Capacitors with a flexible plate can be used to measure strain or pressure. Industrial pressure transmitters used for process control use pressure-sensing diaphragms, which form a capacitor plate of an oscillator circuit. Capacitors are used as the sensor in condenser microphones, where one plate is moved by air pressure, relative to the fixed position of the other plate. Some accelerometers use MEMS capacitors etched on a chip to measure the magnitude and direction of the acceleration vector. They are used to detect changes in acceleration, in tilt sensors, or to detect free fall, as sensors triggering airbag deployment, and in many other applications. Some fingerprint sensors use capacitors. Additionally, a user can adjust the pitch of a theremin musical instrument by moving their hand since this changes the effective capacitance between the user's hand and the antenna.

Changing the effective area of the plates:

Capacitive touch switches are now used on many consumer electronic products.

Oscillators

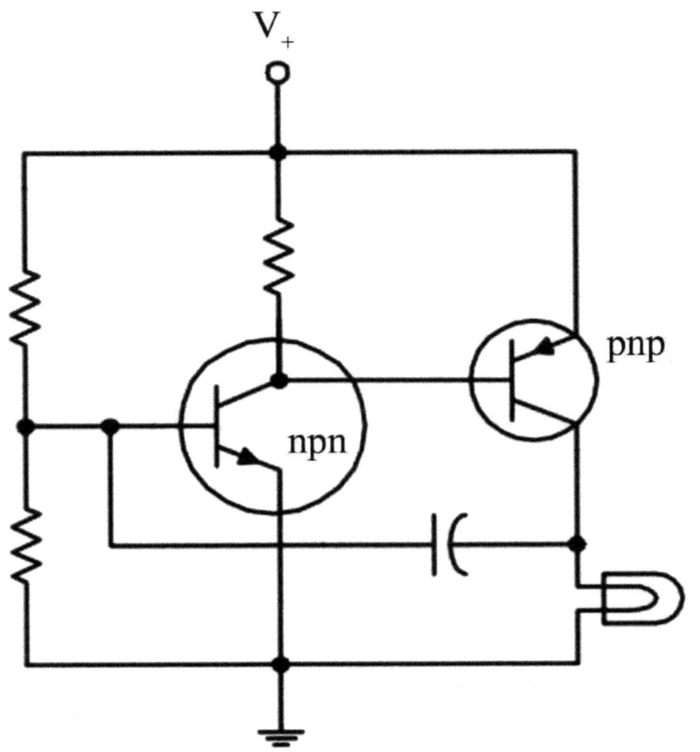

Example of a simple oscillator incorporating a capacitor

A capacitor can possess spring-like qualities in an oscillator circuit. In the image example, a capacitor acts to influence the biasing voltage at the npn transistor's base. The resistance values of the voltage-divider resistors and the capacitance value of the capacitor together control the oscillatory frequency.

Producing Light

A light-emitting capacitor is made from a dielectric that uses phosphorescence to produce light. If one of the conductive plates is made with a transparent material, the light is visible. Light-emitting capacitors are used in the construction of electroluminescent panels, for applications such as backlighting for laptop computers. In this case, the entire panel is a capacitor used for the purpose of generating light.

Hazards and Safety

The hazards posed by a capacitor are usually determined, foremost, by the amount of energy stored, which is the cause of things like electrical burns or heart fibrillation. Factors such as voltage and chassis material are of secondary consideration, which are more related to how easily a shock can be initiated rather than how much damage can occur.

Capacitors may retain a charge long after power is removed from a circuit; this charge can cause dangerous or even potentially fatal shocks or damage connected equipment. For example, even a seemingly innocuous device such as a disposable-camera flash unit, powered by a 1.5 volt AA battery, has a capacitor which may contain over 15 joules of energy and be charged to over 300 volts. This is easily capable of delivering a shock. Service procedures for electronic devices usually include instructions to discharge large or high-voltage capacitors, for instance using a Brinkley stick. Capacitors may also have built-in discharge resistors to dissipate stored energy to a safe level within a few seconds after power is removed. High-voltage capacitors are stored with the terminals shorted, as protection from potentially dangerous voltages due to dielectric absorption or from transient voltages the capacitor may pick up from static charges or passing weather events.

Some old, large oil-filled paper or plastic film capacitors contain polychlorinated biphenyls (PCBs). It is known that waste PCBs can leak into groundwater under landfills. Capacitors containing PCB were labelled as containing "Askarel" and several other trade names. PCB-filled paper capacitors are found in very old (pre-1975) fluorescent lamp ballasts, and other applications.

Capacitors may catastrophically fail when subjected to voltages or currents beyond their rating, or as they reach their normal end of life. Dielectric or metal interconnection failures may create arcing that vaporizes the dielectric fluid, resulting in case bulging, rupture, or even an explosion. Capacitors used in RF or sustained high-current applications can overheat, especially in the center of the capacitor rolls. Capacitors used within high-energy capacitor banks can violently explode when a short in one capacitor causes sudden dumping of energy stored in the rest of the bank into the failing unit. High voltage vacuum capacitors can generate soft X-rays even during normal operation. Proper containment, fusing, and preventive maintenance can help to minimize these hazards.

High-voltage capacitors may benefit from a pre-charge to limit in-rush currents at power-up of high voltage direct current (HVDC) circuits. This extends the life of the component and may mitigate high-voltage hazards.

Swollen electrolytic capacitors – the special design of the capacitor tops allows them to vent instead of bursting violently.

This high-energy capacitor from a defibrillator has a resistor connected between the terminals for safety, to dissipate stored energy.

Catastrophic failure of a capacitor has scattered fragments of paper and metallic foil

CAPACITANCE METER

A capacitance meter is a piece of electronic test equipment used to measure capacitance, mainly of discrete capacitors. Depending on the sophistication of the meter, it may display the capacitance only, or it may also measure a number of other parameters such as leakage, equivalent series resistance (ESR), and inductance. For most purposes and in most cases the capacitor must be disconnected from circuit; ESR can usually be measured in circuit.

Simple Checks without a True Capacitance Meter

Some checks can be made without a specialized instrument, particularly on aluminum electrolytic capacitors which tend to be of high capacitance and to be subject to poor leakage. A multimeter in a resistance range can detect a short-circuited capacitor (very low resistance) or one with very high leakage (high resistance, but lower than it should be; an ideal capacitor has infinite DC resistance). A crude idea of the capacitance can be derived with an analog multimeter in a high resistance range by observing the needle when first connected; current will flow to charge the capacitor and the needle will “kick” from infinite indicated resistance to a relatively low value, and then drift up to infinity. The amplitude of the kick is an indication of capacitance. Interpreting results requires some experience, or comparison with a good capacitor, and depends upon the particular meter and range used.

Simple and Non-Bridge Meters

Many DVMs (digital volt meters) have a capacitance-measuring function. These usually operate by charging and discharging the capacitor under test with a known current and measuring the rate of rise of the resulting voltage; the slower the rate of rise, the larger the capacitance. DVMs can usually measure capacitance from nanofarads to a few hundred microfarads, but wider ranges are not unusual.

- Measurement the capacity of rotary capacitor with digital multimeter

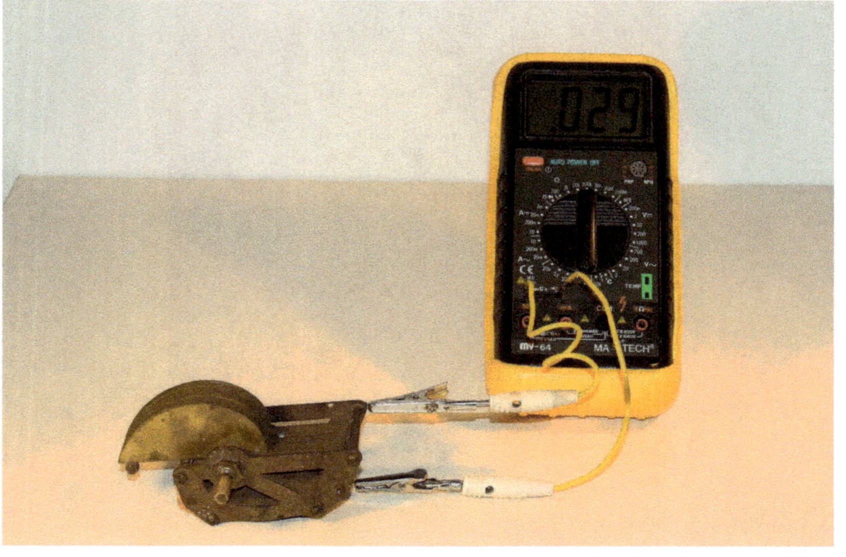

$C_{min} = 29$ pF

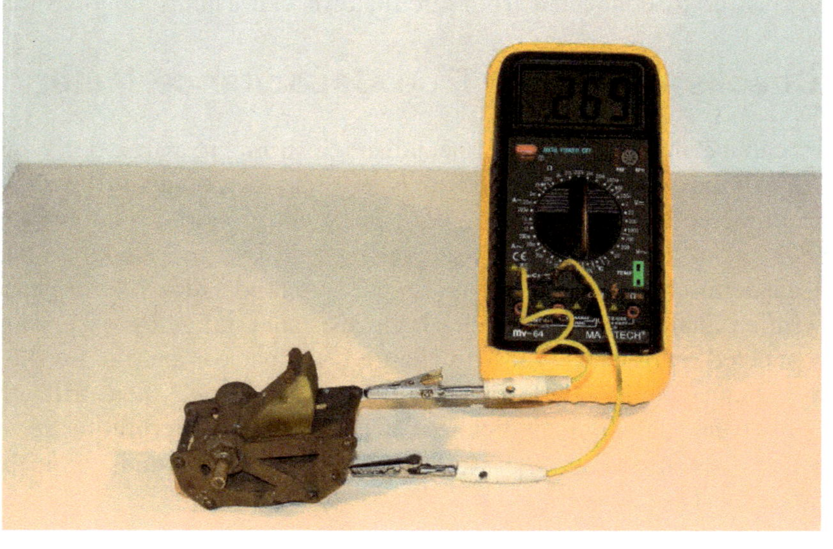

C = 269 pF

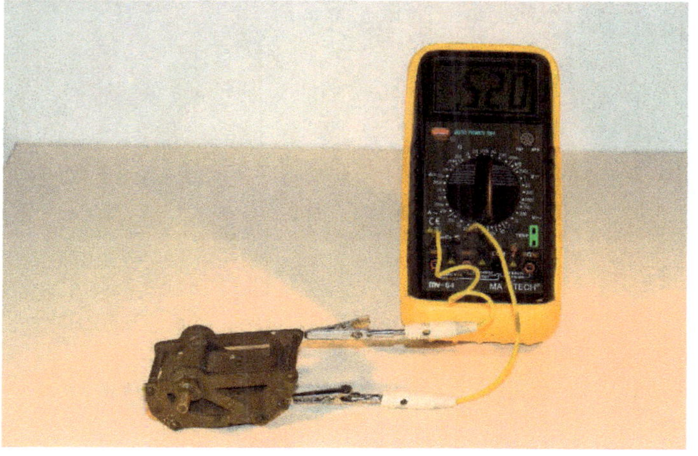

C_{max} = 520 pF

It is also possible to measure capacitance by passing a known high-frequency alternating current through the device under test and measuring the resulting voltage across it (does not work for polarized capacitors).

When troubleshooting circuit problems, a few problems are intermittent or only show up with the working voltage applied, and are not revealed by measurements with equipment, however sophisticated, which uses low test voltages. Some problems are revealed by using a "freezer" spray and observing the effect on circuit operation. Ultimately, in difficult cases routine replacement of capacitors (relatively cheap components) is easier than arranging measurements of all relevant parameters in working conditions.

Some more specialized instruments measure capacitance over a wide range using the techniques described above, and can also measure other parameters. Low stray and parasitic capacitance can be measured if a low enough range is available. Leakage current is measured by applying a direct voltage and measuring the current in the normal way.

Bridges

More sophisticated instruments use other techniques such as inserting the capacitor-under-test into a bridge circuit. By varying the values of the other legs in the bridge (so as to bring the bridge into balance), the value of the unknown capacitor is determined. This method of indirect use of measuring capacitance ensures greater precision. The bridge usually can also measure series resistance and inductance. Through the use of Kelvin connections and other careful design techniques, these instruments can usually measure capacitors over a range from picofarads to farads. Combined LCR meters that can measure inductance, resistance, and capacitance are available.

An Andeen-Hagerling 2700A capacitance bridge

Bridge circuits do not themselves measure leakage current, but a DC bias voltage can be applied and the leakage measured directly.

Modern bridge instruments usually include a digital display and, where relevant, some sort of go/no go testing to allow simple automated use in a production environment. As with all modern instruments, bridges can be interfaced to computer and other equipment to export readings and allow external control.

CAPACITOR PLAGUE

The capacitor plague was a problem related to a higher-than-expected failure rate of non-solid aluminum electrolytic capacitors, between 1999 and 2007, especially those from some Taiwanese manufacturers, due to faulty electrolyte composition that caused corrosion accompanied by gas generation, often rupturing the case of the capacitor from the build-up of pressure.

High failure rates occurred in many well-known brands of electronics, and was particularly evident in motherboards, video cards, and power supplies of personal computers, leading to premature failure of these devices.

Failed aluminum electrolytic capacitors with open vents in the top of the can, and visible dried electrolyte residue (reddish-brown color).

History

First Announcements

Faulty capacitors have been a problem since capacitors' initial development, but the first flawed capacitors linked to Taiwanese raw material problems were reported by the specialist magazine Passive Component Industry in September 2002. Shortly thereafter, two mainstream electronics journals reported the discovery of widespread prematurely failing capacitors, from Taiwanese manufacturers, in motherboards.

These publications informed engineers and other technically interested specialists, but did not give the issue wider public exposure. That changed when Carey Holzman published his experiences about "leaking capacitors" in the overclocking performance community.

Public Attention

Results of fire on a printed circuit board, caused by leaked electrolyte which short-circuited conductors carrying power

The news from the Holzman publication spread quickly on the Internet and in newspapers, partly due to the spectacular images of the failures — bulging or burst cans, expelled sealing rubber and leaking electrolyte on countless circuit boards. A lot of PC users were affected, and caused an avalanche of reports and comments on thousands of blogs and other web communities.

The quick spread of the news also resulted in many misinformed users and blogs posting pictures of capacitors that had failed due to reasons other than faulty electrolyte.

Prevalence

Most of the affected capacitors were produced from 1999 to 2003 and failed between 2002 and 2005. Problems with capacitors produced with an incorrectly formulated electrolyte have affected equipment manufactured up to at least 2007.

Major vendors of motherboards such as Abit, IBM, Dell, Apple, HP, and Intel were affected by capacitors with faulty electrolyte.

In 2005, Dell spent some US$420 million replacing motherboards outright and on the logistics of determining whether a system was in need of replacement.

Many other equipment manufacturers unknowingly assembled and sold boards with faulty capacitors, and as a result the effect of the capacitor plague could be seen in all kinds of devices worldwide.

Because not all manufactures had offered recalls or repairs, do-it-yourself repair instructions were written and can be found on the Internet.

Responsibility

In the November/December 2002 issue of Passive Component Industry, following its initial story about defective electrolyte, reported that some large Taiwanese manufacturers of electrolytic capacitors were denying responsibility for defective products.

While industrial customers confirmed the failures, they were not able to trace the source of the faulty components. The defective capacitors were marked with previously unknown brands such as "Tayeh", "Choyo", or "Chhsi". The marks were not easily linked to familiar companies or product brands. Failed e-caps with well known brands may have had failures not related to defective electrolyte.

The motherboard manufacturer ABIT Computer Corp. was the only such manufacturer that publicly admitted to defective capacitors obtained from Taiwan capacitor makers being used in its products. However, the company would not reveal the name of the capacitor maker that supplied the tainted products.

End of the Plague

With the first publicized press releases about the widespread problem with premature failures of Taiwanese electrolytic capacitors appearing in September 2002, it might be assumed that by mid-2003 the affected capacitor manufacturers would have changed their production process and used a "correct" electrolyte mixture. Assuming a typically shortened life span of about 1.5 to 3 years for the faulty capacitors, the failures should have remained common from mid-2003 to the start of 2005, then tapered off through mid-2006. The last of the bad capacitors should have failed by 2007. Commentators on the Internet often predicted the year 2007 would be the end point for "bad capacitors".

Date of Manufacture Code

Many manufacturers use a 2-character abbreviation according to the IEC 60062 standard, to code the date of production (date code) of electrolytic capacitors:

- First character: Year of production, M = 2000, N = 2001, P = 2002, R = 2003, S = 2004, T = 2005, U = 2006, V = 2007, W = 2008, X = 2009, A = 2010, B = 2011, C = 2012, D = 2013, E = 2014, F = 2015, G = 2016
- Second character: Month of production, 1 to 9 = Jan. to Sept., O = Oct., N = Nov., D = Dec

Example: V8 = August 2007

Symptoms

Common Characteristics

The non-solid aluminum electrolytic capacitors with improperly formulated electrolyte mostly belonged to the so-called "low ESR", "low impedance", or "high ripple current" e-cap series. The advantage of e-caps using an electrolyte composed of 70% water or more is, in particular, a low ESR, which allows a higher ripple current, as well as the use of water making them less expensive to produce, water being the least costly material in a capacitor.

Comparison of aluminum e-caps with different non-solid electrolytes				
Electrolyte	**Manufacturer Series, Type**	**Dimensions D x L (mm)**	**Max. ESR at 100 kHz, 20 °C (mΩ)**	**Max. Ripple current at 85/105 °C (mA)**
Non-solid organic electrolyte	Vishay 036 RSP, 100µF/10V	5x11	1000	160
Non-solid, Ethylene glycol, boric acid (borax) electrolyte	NCC SMQ, 100µF/10V	5x11	900	180
Non-solid water-based electrolyte	Rubycon ZL, 100µF/10V	5x11	300	250

Premature Failure

All electrolytic capacitors with non-solid electrolyte age over time, due to evaporation of the electrolyte. The capacitance usually decreases and the equivalent series resistance (ESR) usually increases. The normal lifespan of a non-solid electrolytic capacitor of consumer quality, typically rated at 2000 h/85 °C and operating at 40 °C, is roughly 6 years. It can be more than 10 years for a 1000 h/105 °C capacitor operating at 40 °C. Electrolytic capacitors that operate at a lower temperature can have a considerably longer lifespan.

The capacitance should normally degrade to as low as 70% of the rated value, and the ESR increase to twice the rated value, over the normal life span of the component, before it should be considered as a degradation failure. The life of an electrolytic capacitor with defective electrolyte can be as little as two years. The capacitor may fail prematurely after reaching approximately 30% to 50% of its expected lifetime.

Electrical Symptoms

The electrical characteristics of a failed electrolytic capacitor with an open vent are the following:

- capacitance value decreases to below the rated value
- ESR increases to very high values.

Electrolytic capacitors with an open vent are in the process of drying out, regardless of whether they have good or bad electrolyte. They always show low capacitance values and very high ohmic ESR values. Dry e-caps are therefore electrically useless.

E-caps can fail without any visible symptoms. Since the electrical characteristics of electrolytic capacitors are the reason for their use, these parameters must be tested with instruments to definitively decide if the devices have failed. But even if the electrical parameters are out of their specifications, the assignment of failure to the electrolyte-problem is not a certainty.

Non-solid aluminum electrolytic capacitors without visible symptoms, which have improperly formulated electrolyte, typically show two electrical symptoms:

- relatively high and fluctuating leakage current
- increased capacitance value, up to twice the rated value, which fluctuates after heating and cooling of the capacitor body

Visible Symptoms

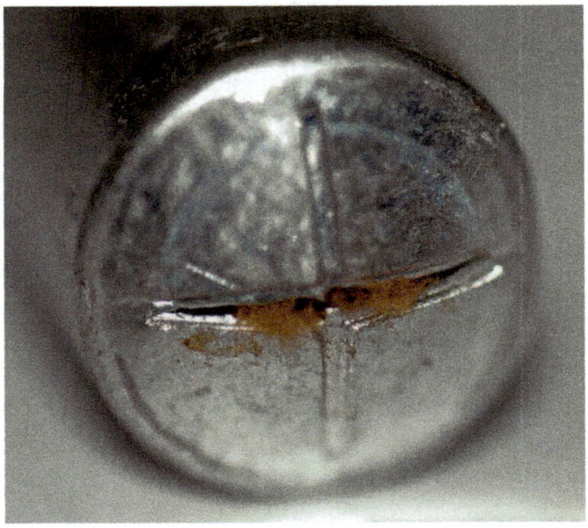

Closeup of a broken electrolytic capacitor vent and dried electrolyte residue

When examining a failed electronic device, the failed capacitors can easily be recognized by clearly visible symptoms that include the following:

- Bulging of the vent on top of the capacitor. (The "vent" is stamped into the top of the casing of a can-shaped capacitor, forming a seam that is meant to split to relieve pressure build-up inside, preventing an explosion.)

- Broken or cracked vent, often accompanied with visible crusty rust-like brown or red dried electrolyte deposits.
- Capacitor casing sitting crooked on the circuit board, caused by the bottom rubber plug being pushed out, sometimes with electrolyte having leaked onto the motherboard from the base of the capacitor, visible as dark-brown or black surface deposits on the PCB. The leaked electrolyte can be confused with thick elastic glue sometimes used to secure the capacitors against shock. A dark brown or black crust up the side of a capacitor is invariably glue, not electrolyte. The glue itself is harmless.

Visible Symptoms of Failed Electrolytic Capacitors

Failed Chhsi capacitor with crusty electrolyte buildup on the top

Failed capacitors next to CPU motherboard socket

Failed Tayeh capacitors which have vented subtly through their aluminum tops

Failed electrolytic capacitors with swollen can tops and expelled rubber seals, dates of manufacture "0106" and "0206" (June 2001 and June 2002)

Failed capacitor has exploded and exposed internal elements, and another has partially blown off its casing

Failed Choyo capacitors (black color) which have leaked brownish electrolyte onto the motherboard

Non-Solid Aluminum Electrolytic Capacitors

The first electrolytic capacitor developed was an aluminum electrolytic capacitor with a liquid electrolyte, invented by Charles Pollak in 1896. Modern electrolytic capacitors are based on the same fundamental design. After roughly 120 years of development billions of these inexpensive and reliable capacitors are used in electronic devices.

Basic Construction

Basic construction details of non-solid aluminum electrolytic capacitors

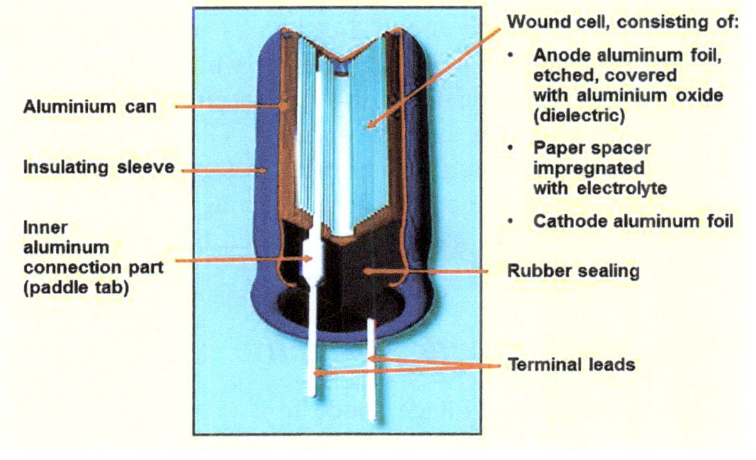

Construction of a typical single-ended aluminum electrolytic capacitor with non-solid electrolyte

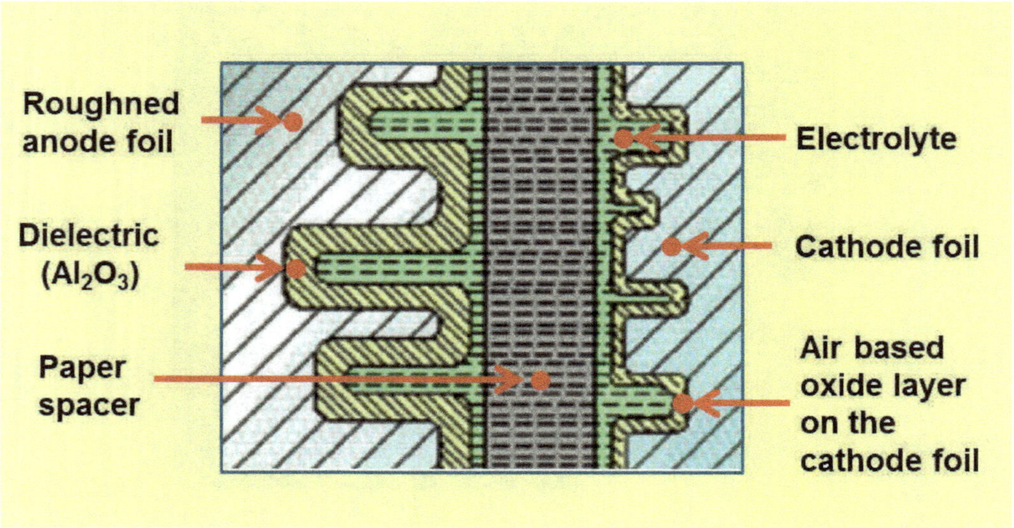

Closeup cross-section diagram of electrolytic capacitor, showing capacitor foils and oxide layers

Aluminum electrolytic capacitors with non-solid electrolyte are generally called "electrolytic capacitors" or "e-caps". The components consist of two strips of aluminum foil, separated by a paper spacer, which is saturated with a liquid or gel-like electrolyte. One of the aluminum foil strips, called the anode, chemically roughened and oxidized in a process called forming, holds a very thin oxide layer on its surface as an electrical insulator serving as the dielectric of the capacitor. The liquid electrolyte, which is the cathode of the capacitor, covers the irregular surface of the oxide layer of the anode perfectly, and makes the increased anode surface effectual, thus increasing the effective capacitance.

A second aluminum foil strip, called the "cathode foil", serves to make electrical contact with the electrolyte. The spacer separates the foil strips to avoid direct metallic contact which would produce a short circuit. Lead wires are attached to both foils which are then rolled with the spacer into a wound cylinder which will fit inside an aluminum case or "can". The winding is impregnated with liquid electrolyte. This provides a reservoir of electrolyte to extend the lifetime of the capacitor. The assembly is inserted into an aluminum can and sealed with a plug. Aluminum electrolytic capacitors with non-solid electrolyte have grooves in the top of the case, forming a vent, which is designed to split open in the event of excessive gas pressure caused by heat, short circuit, or failing electrolyte.

Forming the Aluminum-Oxide Dielectric

View onto the structures of a low-voltage anode foil

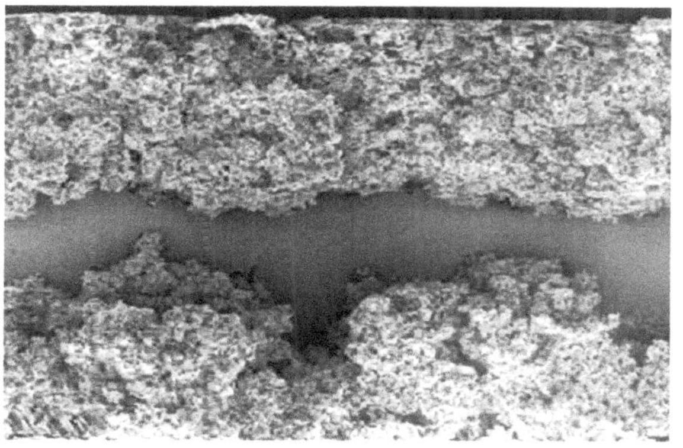

Cross-section side view of etched 10 V low voltage anode foil

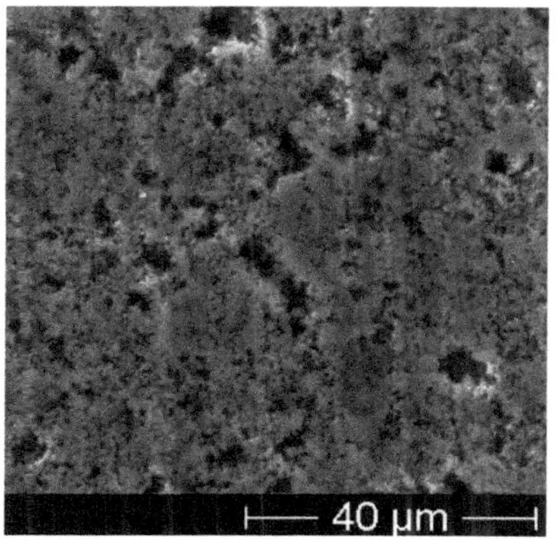

SEM image of the rough anode surface of an unused electrolytic capacitor, showing the openings of pores in the anode

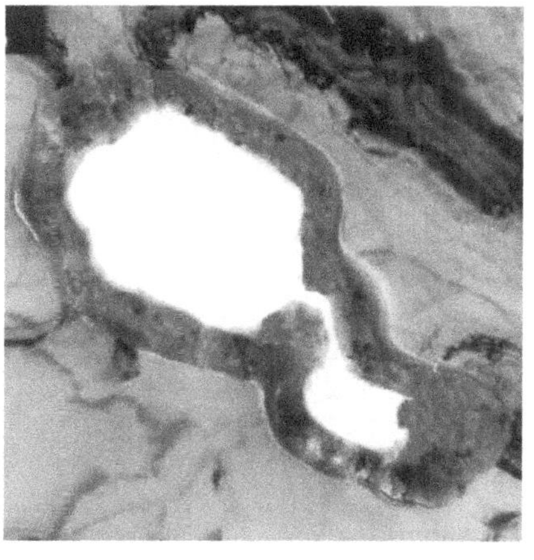

Ultra-thin-cross-section of a etched pore in a low-voltage anode foil, 100,000-fold magnification, light grey: aluminum, dark grey: amorphous aluminum oxide, light: pore, in which the electrolyte is active

The aluminum foil used in non-solid aluminum electrolytic capacitors must have a purity of 99.99%. The foil is roughened by electrochemical etching to enlarge the effective capacitive surface. This etched anode aluminum foil is oxidized (called forming). Forming creates a very thin oxide barrier layer on the anode surface. This oxide layer is electrically insulating and serves as the dielectric of the capacitor. The forming takes place whenever a positive voltage is applied to the anode, and generates an oxide layer whose thickness varies according to the applied voltage. This electrochemical behavior explains the self-healing mechanism of non-solid electrolytic capacitors.

The normal process of oxide formation or self-healing is carried out in two reaction steps. First, a strongly exothermic reaction transforms metallic aluminum (Al) into aluminum hydroxide, $Al(OH)_3$:

$$2\ Al + 6\ H_2O \rightarrow 2\ Al(OH)_3 + 3\ H_2 \uparrow$$

This reaction is accelerated by a high electric field and by high temperatures, and is accompanied by a pressure buildup in the capacitor housing, caused by the released hydrogen gas. The gel-like aluminum hydroxide $Al(OH)_3$ (also called alumina trihydrate (ATH), aluminic hydroxide, aluminum(III) hydroxide, or hydrated alumina) is converted, via a second reaction step (usually slowly over a few hours at room temperature, more rapidly in a few minutes at higher temperatures), into the amorphous or crystalline form of aluminum oxide, Al_2O_3:

$$2\ Al(OH)_3 \rightarrow 2\ AlO(OH) + 2\ H_2O \rightarrow Al_2O_3 + 3\ H_2O$$

This oxide serves as dielectric and also protects the capacitor from the aggressive reactions of metallic aluminum to parts of the electrolyte. One problem of the forming or self-healing processes in non-solid aluminum electrolytics is that of corrosion, the electrolyte having to deliver enough oxygen to generate the oxide layer, with water, corrosive of aluminum, being the most efficient way.

Electrolyte Composition

The name "electrolytic capacitor" derives from the electrolyte, the conductive liquid inside the capacitor. As a liquid it can conform to the etched and porous structure of the anode and the grown oxide layer, and form a "tailor-made" cathode.

From an electrical point of view the electrolyte in an electrolytic capacitor is the actual cathode of the capacitor and must have good electrical conductivity, which is actually ion-conductivity in liquids. But it is also a chemical mixture of solvents with acid or alkali additives, which must be non-corrosive (chemically inert) so that the capacitor, whose inner components are made of aluminum, remains stable over its expected lifetime. In addition to the good conductivity of operating electrolytes, there are other requirements, including chemical stability, chemical compatibility with aluminum, and low cost. The electrolyte should also provide oxygen for the forming processes and self-healing. This diversity of requirements for the liquid electrolyte results in a broad variety of proprietary solutions, with thousands of patented electrolytes.

Up to the mid 1990s electrolytes could be roughly placed into two main groups:

- electrolytes based on ethylene glycol and boric acid. In these so-called glycol or borax electrolytes, an unwanted chemical crystal water reaction occurs: "acid

+ alcohol gives ester + water". These borax electrolytes have been standard in electrolytic capacitors for a long time, and have a water content between 5 and 20%. They work up to a maximum temperature of 85 °C or 105 °C in the voltage range up to 600 V.

- almost anhydrous electrolytes based on organic solvents, such as dimethylformamide (DMF), dimethylacetamide (DMA), or γ-butyrolactone (GBL). These capacitors with organic-solvent electrolytes are suitable for temperatures ranging up to 105 °C, 125 °C, or 150 °C; have low leakage current values; and have very good long-term behavior.

It was known that water is a very good solvent for low ohmic electrolytes. However, the corrosion problems linked to water hindered, up to that time, the use of it in amounts larger than 20% of the electrolyte, the water-driven corrosion using the above-mentioned electrolytes being kept under control with chemical inhibitors that stabilize the oxide layer.

Water-Based Electrolyte Capacitors

In the 1990s a third class of electrolytes was developed by Japanese researchers.

- Water-based electrolytes, with up to 70% water, are relatively inexpensive and boast desirable characteristics such as low ESR and higher voltage handling. These electrolytic capacitors are typically labeled "low-impedance", "low-ESR", or "high-ripple-current" with voltage ratings up to 100 V, for low-cost mass-market applications.
- Despite these advantages, researchers faced several challenges during development of water-based electrolytic capacitors.
- Many of the poorly designed capacitors made it to mass market. The capacitor plague is due to faulty electrolytes of this type.

Development of a Water-Based Electrolyte

At the beginning of the 1990s, some Japanese manufacturers started the development of a new, low-ohmic water-based class of electrolytes. Water, with its high permittivity of ε = 81, is a powerful solvent for electrolytes, and possesses high solubility for conductivity-enhancing concentrations of salt ions, resulting in significantly improved conductivity compared to electrolytes with organic solvents like GBL. But water will react quite aggressively and even violently with unprotected aluminum, converting metallic aluminum (Al) into aluminum hydroxide ($Al(OH)_3$), via a highly exothermic reaction that gives off heat, causing gas expansion that can lead to an explosion of the capacitor. Therefore, the main problem in the development of water-based electrolytes is achieving long-term stability by hindering the corrosive action of water on aluminum.

Normally the anode foil is covered by the dielectric aluminum oxide (Al_2O_3) layer, which protects the base aluminum metal against the aggressiveness of aqueous alkali solutions. However, some impurities or weak points in the oxide layer offer the possibility for water-driven anodic corrosion that forms aluminum hydroxide ($Al(OH)_3$). In e-caps using an alkaline electrolyte this aluminum hydroxide will not be transformed into the desired stable form of aluminum oxide. The weak point remains and the anodic corrosion is ongoing. This corrosive process can be interrupted by protective substances in the electrolyte known as inhibitors or passivators. Inhibitors—such as chromates, phosphates, silicates, nitrates, fluorides, benzoates, soluble oils, and certain other chemicals—can reduce the anodic and cathodic corrosion reactions. However, if inhibitors are used in an insufficient amount, they tend to increase pitting.

The Water Problem in Non-Solid Aluminum Electrolytic Capacitors

The aluminum oxide layer in the electrolytic capacitor is resistant to chemical attacks, as long as the pH value of the electrolyte is in the range of pH 4.5 to 8.5. However, the pH value of the electrolyte is ideally about 7 (neutral); and measurements carried out as early as the 1970s have shown that leakage current is increased, due to chemically induced defects, when the pH value deviates from this ideal value. It is known that water is extremely corrosive to pure aluminum and introduces chemical defects. It is further known that unprotected aluminum oxide dielectrics can be slightly dissolved by alkaline electrolytes, weakening the oxide layer.

The fundamental issue of water-containing-electrolyte systems lies in the control of aggressiveness of the water towards metallic aluminum. This issue has dominated the development of electrolytic capacitors over many decades. The first commercially used electrolytes in the mid-twentieth century were mixtures of ethylene glycol and boric acid. But even these glycol electrolytes had an unwanted chemical water-crystal reaction, according to the scheme: "acid + alcohol" → "ester + water". Thus, even in the first apparently water-free electrolytes, esterification reactions could generate a water content of up to 20 percent. These electrolytes had a voltage-dependent life span, because at higher voltages the leakage current based on the aggressiveness of the water would increase exponentially; and the associated increased consumption of electrolyte would lead to a faster drying out. Otherwise, the electrolyte has to deliver the oxygen for self-healing processes, and water is the best chemical substance to do that.

Water-Driven Corrosion: Aluminum Hydroxide

Attempt at a pictorial representation of the formation of aluminum hydroxide in a pore of a roughened electrolytic capacitor anode foil

It is known that the "normal" course of building a stable aluminum oxide layer by the transformation of aluminum, through the intermediate step of aluminum hydroxide, can be interrupted by an excessively alkaline or basic electrolyte. For example, alkaline disruption of the chemistry of this reaction results instead in the following reaction:

$$2\ Al\ (s) + 2\ NaOH\ (aq) + 6\ H_2O \rightarrow 2\ Na^+\ (aq) + 2[Al(OH)_4]^-\ (s) + 3\ H_2\ (g)$$

In this case, it may happen that the hydroxide formed in the first step becomes mechanically detached from the metallic aluminum surface and will not be transformed into the desired stable form of aluminum oxide. The initial self-healing process for building a new oxide layer is prevented by a defect or a weak dielectric point, and generated hydrogen gas escapes into the capacitor. Then, at the weak point, further formation of aluminum hydroxide is started, and prevented from converting into stable aluminum oxide. The self-healing of the oxide layer inside the electrolytic capacitor can't take place. However, reactions do not come to a standstill, as more and more hydroxide grows in the pores of the anode foil, and the first reaction step produces more and more hydrogen gas in the can, increasing the pressure.

Scanning electron microscope (SEM) images of different forms of aluminum hydroxide out of failed electrolytic capacitors.

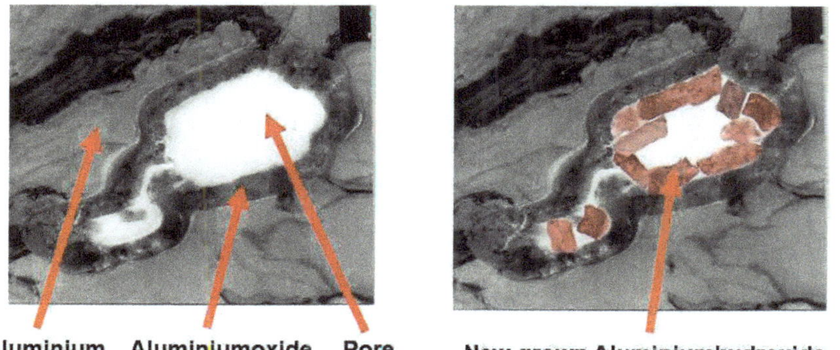

Polished intersection of an anode with a pore covered with dielectric layer

(photoretouching to show the principle)

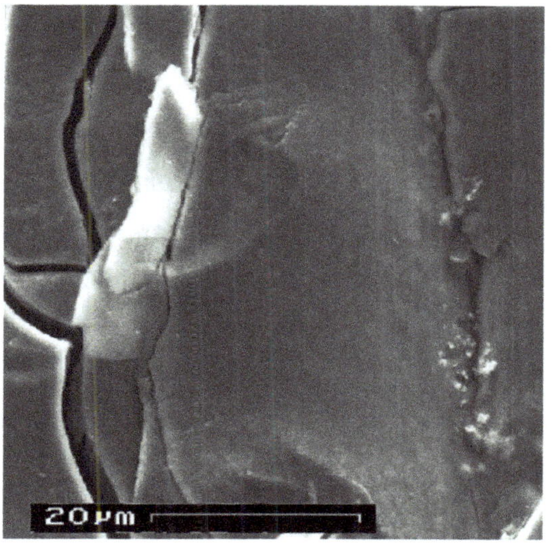

Anode surface with grown plaques of aluminum hydroxide

Anode surface with grown beads of aluminum hydroxide

Manufacturing for Market

The Japanese manufacturer Rubycon became a leader in the development of new water-based electrolyte systems with enhanced conductivity in the late 1990s. After several years of development, researchers led by Shigeru Uzawa had found a mixture of inhibitors that suppressed the aluminum hydration. In 1998, Rubycon announced two series, ZL and ZA, of the first production capacitors using an electrolyte with a water content of about 40%, which were suitable for temperatures ranging from −40 °C (−40 °F; 233 K) to 105 °C (221 °F; 378 K). Later, electrolytes were developed to work with water of up to 70% by weight. Other manufacturers, such as NCC, Nichicon, and Elna followed with their own products a short time later.

The improved conductivity of the new electrolyte can be seen by comparing two capacitors, both of which have a nominal capacitance of 1000 μF at 16 V rated voltage, in a package with a diameter of 10 mm and a height of 20 mm. The capacitors of the Rubycon YXG series are provided with an electrolyte based on an organic solvent and can attain an impedance of 46 mΩ when loaded with a ripple current of 1400 mA. ZL series capacitors with the new water-based electrolyte can attain an impedance of 23 mΩ with a ripple current of 1820 mA, an overall improvement of 30%.

The new type of capacitor was called Low-ESR or Low-Impedance, Ultra-Low-Impedance or High-Ripple Current series in the data sheets. The highly competitive market in digital data technology and high-efficiency power supplies rapidly adopted these new components because of their improved performance. Furthermore, by improving the conductivity of the electrolyte, capacitors not only can withstand a higher ripple current rating, they are much cheaper to produce since water is much cheaper than other solvents. Better performance and low cost drove widespread adoption of the new capacitors for high volume products such as PCs, LCD screens, and power supplies.

Investigation

Implications of Industrial Espionage

Industrial espionage was implicated in the capacitor plague, in connection with the theft of an electrolyte formula. A materials scientist working for Rubycon in Japan left the company, taking the secret water-based electrolyte formula for Rubycon's ZA and ZL series capacitors, and began working for a Chinese company. The scientist then developed a copy of this electrolyte. Then, some staff members who defected from the Chinese company copied an incomplete version of the formula and began to market it to many of the aluminum electrolytic manufacturers in Taiwan, undercutting the prices of the Japanese manufacturers. This incomplete electrolyte lacked important proprietary ingredients which were essential to the long-term stability of the capacitors and was unstable when packaged in a finished aluminum capacitor. This faulty electrolyte allowed the unimpeded formation of hydroxide and produced hydrogen gas.

There are no known public court proceedings related to alleged theft of electrolyte formulas. However, one independent laboratory analysis of defective capacitors has shown that many of the premature failures appear to be associated with high water content and missing inhibitors in the electrolyte, as described below.

Incomplete Electrolyte Formula

Unimpeded formation of hydroxide (hydration) and associated hydrogen gas production, occurring during "capacitor plague" or "bad capacitors" incidents involving the failure of large numbers of aluminum electrolytic capacitors, has been demonstrated by two researchers at the University of Maryland who analyzed the failed capacitors.

The two scientists initially determined, by ion chromatography and mass spectrometry, that there was hydrogen gas present in failed capacitors, leading to bulging of the capacitor's case or bursting of the vent. Thus it was proved that the oxidation takes place in accordance with the first step of aluminum oxide formation.

Because it has been customary in electrolytic capacitors to bind the excess hydrogen with the help of reducing or depolarizing compounds, such as aromatic nitrogen compounds or amines, to relieve the resulting pressure, the researchers then searched for compounds of this type. Although the analysis methods were very sensitive in detecting such pressure-relieving compounds, no traces of such agents were found within the failed capacitors.

In capacitors in which the internal pressure build-up was so great that the capacitor case was already bulging but the vent had not opened yet, the pH value of the electrolyte could be measured. The electrolyte of the faulty Taiwanese capacitors was alkaline, with a pH of between 7 and 8. Good comparable Japanese capacitors had an electrolyte that was acidic, with a pH of around 4. As it is known that aluminum can be dissolved by alkaline liquids, but not that which is mildly acidic, an energy dispersive X-ray spectroscopy (EDX or EDS) fingerprint analysis of the electrolyte of the faulty capacitors was made, which detected dissolved aluminum in the electrolyte.

To protect the metallic aluminum against the aggressiveness of the water, some phosphate compounds, known as inhibitors or passivators, can be used to produce long-term stable capacitors with high-aqueous electrolytes. Phosphate compounds are mentioned in patents regarding electrolytic capacitors with aqueous electrolytic systems. Since phosphate ions were missing and the electrolyte was also alkaline in the investigated Taiwanese electrolytes, the capacitor evidently lacked any protection against water damage, and the formation of more-stable alumina oxides was inhibited. Therefore, only aluminum hydroxide was generated.

The results of chemical analysis were confirmed by measuring electrical capacitance and leakage current in a long-term test lasting 56 days. Due to the chemical corrosion, the oxide layer of these capacitors had been weakened, so after a short time the capacitance and the leakage current increased briefly, before dropping abruptly when gas pressure opened the vent. The report of Hillman and Helmold proved that the cause of the failed capacitors was a faulty electrolyte mixture used by the Taiwanese manufacturers, which lacked the necessary chemical ingredients to ensure the correct pH of the electrolyte over time, for long-term stability of the electrolytic capacitors. Their further conclusion, that the electrolyte with its alkaline pH value had the fatal flaw of a continual buildup of hydroxide without its being converted into the stable oxide, was verified on the surface of the anode foil both photographically and with an EDX-fingerprint analysis of the chemical components.

REFERENCES

1. Alexander, Charles; Sadiku, Matthew. Fundamentals of Electric Circuits (3 ed.). McGraw-Hill. p. 206.
2. Duff, Wilmer (1908–1916). A Text-Book of Physics (4 ed.). Philadelphia: P. Blakiston's Son & Co. p. 361. Retrieved 1 December 2016.
3. Bird, John (2010). Electrical and Electronic Principles and Technology. Routledge. pp. 63–76. ISBN 9780080890562. Retrieved 2013-03-17.
4. Floyd, Thomas (1984–2005). Electronic Devices (7 ed.). Upper Saddle River, NJ: Pearson Education. p. 10. ISBN 0-13-127827-4.
5. Williams, Henry Smith. "A History of Science Volume II, Part VI: The Leyden Jar Discovered". Retrieved 2013-03-17.
6. Keithley, Joseph F. (1999). The Story of Electrical and Magnetic Measurements: From 500 BC to the 1940s. John Wiley & Sons. p. 23. ISBN 9780780311930. Retrieved 2013-03-17.
7. Houston, Edwin J. (1905). Electricity in Every-day Life. P. F. Collier & Son. p. 71. Retrieved 2013-03-17.
8. Isaacson, Walter (2003). Benjamin Franklin: An American Life. Simon and Schuster. p. 136. ISBN 9780743260848. Retrieved 2013-03-17.
9. Franklin, Benjamin (1749-04-29). "Experiments & Observations on Electricity: Letter IV to Peter Collinson" (PDF). p. 28. Retrieved 2009-08-09.
10. Morse, Robert A. (September 2004). "Franklin and Electrostatics—Ben Franklin

as my Lab Partner" (PDF). Wright Center for Science Education. Tufts University. p. 23. Retrieved 2009-08-10. After Volta's discovery of the electrochemical cell in 1800, the term was then applied to a group of electrochemical cells

11. "eFunda: Glossary: Units: Electric Capacitance: Jar". eFunda. Retrieved 2013-03-17.
12. "Sketch of Alessandro Volta". The Popular Science Monthly. New York: Bonnier Corporation: 118–119. May 1892. ISSN 0161-7370.
13. British Engineering Standards Association, British Standard Glossary of Terms in Electrical Engineering, C. Lockwood & son, 1926
14. Ho, J.; Jow, R.; Boggs, S. (Jan 2010). "Historical Introduction to Capacitor Technology" (PDF). IEEE Elect. Insul. Mag. IEEE. 26 (1): 20–25. doi:10.1109/mei.2010.5383924.
15. US 2800616, Becker, H.I., "Low voltage electrolytic capacitor", issued 1957-07-23
16. A brief history of supercapacitors AUTUMN 2007 Batteries & Energy Storage Technology
17. Hammond, Percy (1964). Electromagnetism for Engineers: An Introductory Course. The Commonwealth and International Library of Science, Technology, Engineering and Liberal Studies. Applied Electricity and Electronics Division. 3. Pergamon Press. pp. 44–45.
18. "Capacitor charging and discharging". All About Circuits. Retrieved 2009-02-19.
19. Pillai, K. P. P. (1970). "Fringing field of finite parallel-plate capacitors". Proceedings of the Institution of Electrical Engineers. 117 (6): 1201–1204. doi:10.1049/piee.1970.0232.
20. Pai, S. T.; Qi Zhang (1995). Introduction to High Power Pulse Technology. Advanced Series in Electrical and Computer Engineering. 10. World Scientific. ISBN 9789810217143. Retrieved 2013-03-17.
21. Dyer, Stephen A. (2004). Wiley Survey of Instrumentation and Measurement. John Wiley & Sons. p. 397. ISBN 9780471221654. Retrieved 2013-03-17.
22. Scherz, Paul (2006). Practical Electronics for Inventors (2nd ed.). McGraw Hill Professional. p. 100. ISBN 9780071776448. Retrieved 2013-03-17.
23. Bird, John (2007). Electrical Circuit Theory and Technology. Routledge. p. 501. ISBN 9780750681391. Retrieved 2013-03-17.
24. "Ceramic Capacitor Aging Made Simple". Johanson Dielectrics. 2012-05-21. Retrieved 2013-03-17.
25. "The Effect of Reversal on Capacitor Life" (PDF). Engineering Bulletin 96-004. Sorrento Electronics. November 2003. Archived from the original (PDF) on 2014-07-14. Retrieved 2013-03-17.
26. Kaiser, Cletus J. (1993) The Capacitor Handbook. Springer
27. Electronics. McGraw-Hill 1960 p. 90
28. Xenon Strobe and Flash Safety Hints. donklipstein.com. May 29, 2006
29. Prutchi, David (2012). Exploring Quantum Physics through Hands-on Projects.

John Wiley and Sons. p. 10. ISBN 1118170709.

30. Dixit, J. B. and Yadav, Amit (2010) Electrical Power Quality. University Science Press. p. 63. ISBN 9380386745.
31. Winburn (1990) Practical Laser Safety Second Edition. Marcel-Dekker Inc. p. 189. ISBN 0824782402.
32. Guinta, Steve. "Ask The Applications Engineer – 21". Analog Devices. Retrieved 2013-03-17.
33. Carlos Paz de Araujo, Ramamoorthy Ramesh, George W Taylor (Editors) (2001). Science and Technology of Integrated Ferroelectrics: Selected Papers from Eleven Years of the Proceedings of the International Symposium on Integrated Ferroelectrics. CRC Press. Figure 2, p. 504. ISBN 90-5699-704-1.
34. Solomon Musikant (1991). What Every Engineer Should Know about Ceramics. CRC Press. Figure 3.9, p. 43. ISBN 0-8247-8498-7.
35. Yasuo Cho (2005). Scanning Nonlinear Dielectric Microscope (in Polar Oxides; R Waser, U Böttger & S Tiedke, editors ed.). Wiley-VCH. Chapter 16. ISBN 3-527-40532-1.
36. Simon M. Sze; Kwok K. Ng (2006). Physics of Semiconductor Devices (3rd ed.). Wiley. Figure 25, p. 121. ISBN 0-470-06830-2.
37. Gabriele Giuliani; Giovanni Vignale (2005). Quantum Theory of the Electron Liquid. Cambridge University Press. p. 111. ISBN 0-521-82112-6.
38. Jørgen Rammer (2007). Quantum Field Theory of Non-equilibrium States. Cambridge University Press. p. 158. ISBN 0-521-87499-8.
39. Horst Czichos; Tetsuya Saito; Leslie Smith (2006). Springer Handbook of Materials Measurement Methods. Springer. p. 475. ISBN 3-540-20785-6.
40. William Coffey; Yu. P. Kalmykov (2006). Fractals, diffusion and relaxation in disordered complex systems..Part A. Wiley. p. 17. ISBN 0-470-04607-4.
41. J. Obrzut, A. Anopchenko and R. Nozaki, "Broadband Permittivity Measurements of High Dielectric Constant Films", Proceedings of the IEEE: Instrumentation and Measurement Technology Conference, 2005, pp. 1350–1353, 16–19 May 2005, Ottawa ISBN 0-7803-8879-8 doi:10.1109/IMTC.2005.1604368
42. Dieter K Schroder (2006). Semiconductor Material and Device Characterization (3rd ed.). Wiley. p. 347 ff. ISBN 0-471-73906-5.
43. Dieter K Schroder (2006). Semiconductor Material and Device Characterization (3rd ed.). Wiley. p. 270 ff. ISBN 0-471-73906-5.
44. Simon M. Sze; Kwok K. Ng (2006). Physics of Semiconductor Devices (3rd ed.). Wiley. p. 217. ISBN 0-470-06830-2.
45. Safa O. Kasap; Peter Capper (2006). Springer Handbook of Electronic and Photonic Materials. Springer. Figure 20.22, p. 425.
46. PY Yu; Manuel Cardona (2001). Fundamentals of Semiconductors (3rd ed.). Springer. p. §6.6 Modulation Spectroscopy. ISBN 3-540-25470-6.
47. Kaplan, Daniel M.; White, Christopher G. Hands-On Electronics: A Practical

Introduction to Analog and Digital Circuits. p. 19.

48. Capacitor MF-MMFD Conversion Chart; Just Radios.
49. Fundamentals of Electronics – Volume 1b – Basic Electricity – Alternating Current; Bureau of Naval Personnel; 1965; page 197.
50. Miller, Charles. Illustrated Guide to the National Electrical Code, p. 445 (Cengage Learning 2011).
51. Alam, Mohammed; Michael H. Azarian; Michael Osterman; Michael Pecht (2010). Effectiveness of embedded capacitors in reducing the number of surface mount capacitors for decoupling applications. Circuit World. 36 (1): 22. doi:10.1108/03056121011015068.
52. Downie, Neil A and Mathilde Pradier, ‹Method and apparatus for monitoring fluid pressure, US Patent 7526961 (2009)
53. D. M. Zogbi (September 2002). Low-ESR Aluminum Electrolytic Failures Linked to Taiwanese Raw Material Problems (PDF). Passive Component Industry. Paumanok Publications. 4 (5): 10, 12, 31. Archived from the original (PDF) on 2015-11-03. Retrieved 2015-11-03.
54. The Capacitor Plague, Posted on 26 November 2010 by PC Tools
55. Sperling, Ed; Soderstrom, Thomas; Holzman, Carey (October 2002). Got Juice?. EE Times.
56. Chiu, Yu-Tzu; Moore, Samuel K (February 2003). Faults & Failures: Leaking capacitors muck up motherboards. IEEE Spectrum. 40 (2): 16–17. doi:10.1109/MSPEC.2003.1176509. ISSN 0018-9235. Retrieved 2014-08-22.
57. Carey Holzman, Overclockers, Capacitors: Not Just For Abit Owners, Motherboards with leaking capacitors, 10/9, 2002,
58. Hales, Paul (5 November 2002). Taiwanese component problems may cause mass recalls. The Inquirer. Retrieved 27 April 2015.
59. Capacitor failures plague motherboard vendors, GEEK, 7 February 2003
60. W. BONOMO, G. HOOPER, D. RICHARDSON, D. ROBERTS, and TH. VAN DE STEEG, Vishay Intertechnology, Failure modes in capacitors,
61. Mainboardhersteller steht für Elko-Ausfall gerade, Heise (in German) (online ed.), DE.
62. Michael Singer, CNET News, Bulging capacitors haunt Dell, 31 October 2005
63. Michael Singer, CNET News, PCs plagued by bad capacitors
64. The guardian technology blog, How a stolen capacitor formula ended up costing Dell $300m
65. Vance, Ashlee (28 June 2010). Suit Over Faulty Computers Highlights Dell›s Decline. The New York Times. Retrieved 2012-03-08.
66. Capacitor Replacement Video Tutorial (HD) (video), Afro tech mods.
67. Repair and bad capacitor information, Capacitor Lab.
68. Liotta, Bettyann (November 2002). Taiwanese Cap Makers Deny Responsibility

(PDF). Passive Component Industry. Paumanok Publications. 4 (6): 6, 8–10. Archived from the original (pdf) on 20 November 2015. Retrieved 2015-11-03.

69. Capacitor plague, identifizierte Hersteller (~identified vendors). Opencircuits.com. 2012-01-10. Retrieved 2014-09-03.
70. Uzawa, Shigeru; Komatsu, Akihiko; Ogáwara, Tetsushi; Rubycon Corporation (2002). Ultra Low Impedance Aluminum Electrolytic Capacitor with Water based Electrolyte. Journal of Reliability Engineering Association of Japan. 24 (4): 276–283. ISSN 0919-2697. Accession number 02A0509168.
71. A. Albertsen, Electrolytic Capacitor Lifetime Estimation (PDF). Retrieved 2014-09-04.
72. Sam G. Parler, Cornell Dubilier, Deriving Life Multipliers for Electrolytic Capacitors
73. The Aluminum Electrolytic Condenser, H. 0. Siegmund, Bell System Technical Journal, v8, 1. January 1229, pp. 41–63
74. A. Güntherschulze, H. Betz, Elektrolytkondensatoren, Verlag Herbert Cram, Berlin, 2. Auflage 1952
75. Motherboard Capacitor Problem Blows Up. Silicon Chip. AU. 2003-05-11. Retrieved 2012-03-07.
76. Blown, Burst and Leaking Motherboard Capacitors - A Serious Problem, PCSTATS, Jan 15 2005
77. Sundoc Bibliothek, Universität Halle, Dissertation, Aluminum anodization,
78. Elna, Principles, 3. Electrolyte, Table 2: An Example of the Composition of the Electrolyte
79. NON-AQUEOUS ELECTROLYTES and THEIR CHARACTERISTICS, FaradNet Electrolytic Capacitors, Part III: Chapter 10
80. K. H. Thiesbürger: Der Elektrolyt-Kondensator. 4. Auflage. Roederstein, Landshut 1991, [OCLC31349250]
81. W. J. Bernard, J. J. Randall Jr., The Reaction between Anodic Aluminum Oxide and Water,
82. Ch. Vargel, M. Jacques, M. P. Schmidt, Corrosion of Aluminium, 2004 Elsevier B.V., ISBN 978-0-08-044495-6
83. Alfonso Berduque, Zongli Dou, Rong Xu, BHC Components Ltd (KEMET), Electrochemical Studies for Aluminium Electrolytic Capacitor Applications: Corrosion Analysis of Aluminium in Ethylene Glycol-Based Electrolytes
84. J.L. Stevens, T. R. Marshall, A.C. Geiculescu m, C.R. Feger, T.F. Strange, Carts USA 2006, The Effects of Electrolyte Composition on the Deformation Characteristics of Wet Aluminum ICD Capacitors, Archived copy (PDF). Archived from the original (PDF) on 14 December 2014. Retrieved 2014-12-14.
85. Bernard, Walter J; Randall Jr, John J (7 April 1961). The Reaction between Anodic Aluminum Oxide and Water (PDF). Journal of The Electrochemical Society. 154 (7): 355–361. doi:10.1149/1.2428230. Archived from the original

(PDF) on 23 December 2014. Retrieved 2015-11-03.

86. Alu Encyclopaedia, oxide layer. Aluinfo.de. Retrieved 2014-09-04
87. J. M. Sanz, J. M. Albella, J. M. Martinez-Duart, On the inhibition of the reaction between anodic aluminum oxide and water
88. Hillman; Helmold (2004). Identification of Missing or Insufficient Electrolyte Constituents in Failed Aluminum Electrolytic Capacitors (PDF), DFR solutions
89. K. H. Thiesbürger: Der Elektrolyt-Kondensator 4th edition, Page 88 to 91, Roederstein, Landshut 1991 (OCLC 313492506)
90. H. Kaesche, Die Korrosion der Metalle - Physikalisch-chemische Prinzipien und aktuelle Probleme, Springer-Verlag, Berlin, 1966, ISBN 978-3-540-51569-2 (1990 edition)
91. Ncc, Ecc. Chemi-con.co.jp. Retrieved 2014-09-04.
92. Nichicon. Nichicon-us.com. Retrieved 2014-09-04.
93. Elna. Elna. Retrieved 2014-09-04.
94. Low-ESR Aluminium Electrolytic Failures Linked to Taiwanese Raw Material Problems (PDF), Molalla, archived from the original (PDF) on 26 April 2012
95. Chang, Jeng-Kuei, Liao, Chi-Min, Chen, Chih-Hsiung, Tsai, Wen-Ta, Effect of electrolyte composition on hydration resistance of anodized aluminum oxide

CHAPTER 4

MOSFET Amplifiers

Chapter Contents

- **Introduction**
- **History**
- **Composition**
- **Operation**
- **Circuit Symbols**
- **Applications**
- **Construction**
- **Operational Amplifier**

In general terms, an amplifier is a device that takes a signal and increases its amplitude. It is part of what is known as an audio playback chain, which typically includes an audio source, such as a CD or record player, and other devices, like pre-amps and equalizers, and ends in a speaker. A MOSFET amplifier is a subcategory of amplifiers that employs metal–oxide–semiconductor field-effect transistor (MOSFET) technology to process digital signals with relatively low power consumption. Once limited to only high-end products, by the early 21st century, MOSFET circuits are the design of choice in more than 99% of microchips manufactured worldwide, including those used in amplifiers.

The idea of using metal-oxide as an insulator in a transistor dates back to the mid-1920s, when the general concept was first proposed and patented in the U.S. by Austro-Hungarian scientist Julius Edgar Lilienfeld. His idea was that metal-oxide layers could control the voltages pushed through a transistor with less of the performance hindrances that plagued existing field-effect transistor (FET) designs. True MOSFET, as it would be recognized today, was developed in the 1960s by Dawon Kahng, who demonstrated the first successful example of an insulated-gate field-effect transistor with gates comprised of metal, oxide, and silicon.

INTRODUCTION

The metal–oxide–semiconductor field-effect transistor (MOSFET, MOS-FET, or MOS FET) is a type of field-effect transistor (FET). It has an insulated gate, whose voltage determines the conductivity of the device. This ability to change conductivity with the amount of applied voltage can be used for amplifying or switching electronic signals. Although FET is sometimes used when referring to MOSFET devices, other types of field-effect transistors also exist.

Although the MOSFET is a four-terminal device with source (S), gate (G), drain (D), and body (B) terminals, the body (or substrate) of the MOSFET is often connected to the source terminal, making it a three-terminal device like other field-effect transistors. Because these two terminals are normally connected to each other (short-circuited) internally, only three terminals appear in electrical diagrams.

The basic principle of the field-effect transistor was first patented by Julius Edgar Lilienfeld in 1925.

The main advantage of a MOSFET over a regular transistor is that it requires very little current to turn on (less than 1mA), while delivering a much higher current to a load (10 to 50A or more).

In *enhancement mode* MOSFETs, a voltage drop across the oxide induces a conducting channel between the source and drain contacts *via* the field effect. The term "enhancement mode" refers to the increase of conductivity with increase in oxide field that adds carriers to the channel, also referred to as the *inversion layer*. The channel can contain electrons (called an nMOSFET or nMOS), or holes (called a pMOSFET or pMOS), opposite in type to the substrate, so nMOS is made with a p-type substrate, and pMOS with an n-type substrate. In the less common *depletion mode* MOSFET, detailed later on, the channel consists of carriers in a surface impurity layer of opposite type to the substrate, and conductivity is decreased by application of a field that depletes carriers from this surface layer.

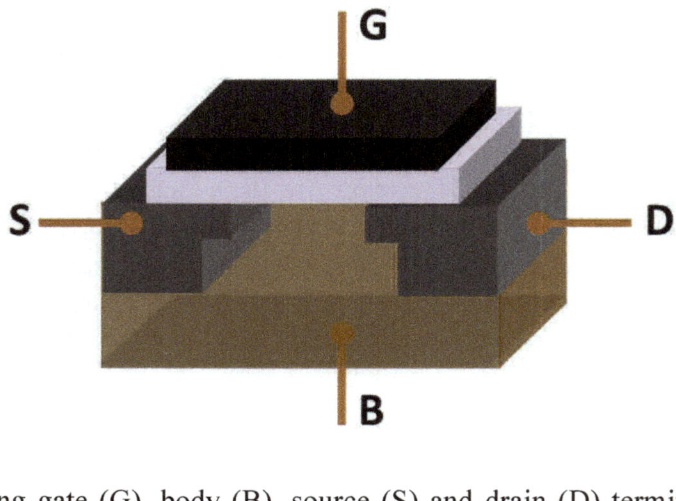

MOSFET showing gate (G), body (B), source (S) and drain (D) terminals. The gate is separated from the body by an insulating layer (white)

Two power MOSFETs in D2PAK surface-mount packages. Operating as switches, each of these components can sustain a blocking voltage of 120 volts in the OFF state, and can conduct a continuous current of 30 amperes in the ON state, dissipating up to about 100 watts and controlling a load of over 2000 watts. A matchstick is pictured for scale

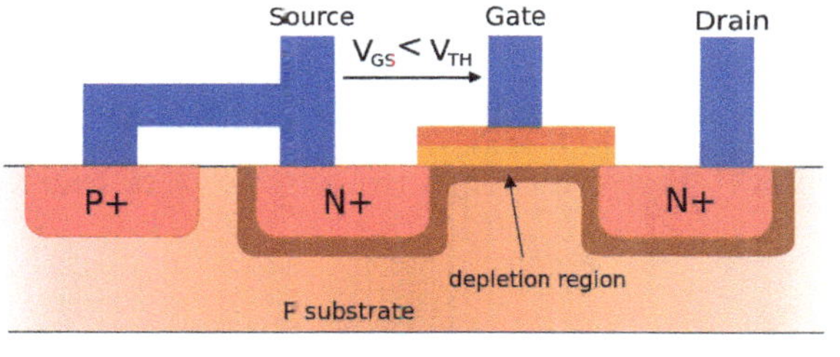

A cross section through an nMOSFET when the gate voltage VGS is below the threshold for making a conductive channel; there is little or no conduction between the terminals drain and source; the switch is off. When the gate is more positive, it attracts electrons, inducing an n-type conductive channel in the substrate below the oxide, which allows electrons to flow between the n-doped terminals; the switch is on.

The "metal" in the name MOSFET is now often a misnomer because the previously metal gate material is now often a layer of polysilicon (polycrystalline silicon). Aluminium had been the gate material until the mid-1970s, when polysilicon became dominant, due to its capability to form self-aligned gates. Metallic gates are regaining popularity, since it is difficult to increase the speed of operation of transistors without metal gates. Likewise, the "oxide" in the name can be a misnomer, as different dielectric materials are used with the aim of obtaining strong channels with smaller applied voltages.

An insulated-gate field-effect transistor or IGFET is a related term almost synonymous with MOSFET. The term may be more inclusive, since many "MOSFETs" use a gate that is not metal, and a gate insulator that is not oxide. Another synonym is MISFET for metal–insulator–semiconductor FET.

The MOSFET is by far the most common transistor in both digital and analog circuits, though the bipolar junction transistor was at one time much more common.

HISTORY

The basic principle of this kind of transistor was first patented by Julius Edgar Lilienfeld in 1925. Twenty five years later, when Bell Telephone attempted to patent the junction transistor, they found Lilienfeld already holding a patent, worded in a way that would include all types of transistors. Bell Labs was able to work out an agreement with Lilienfeld, who was still alive at that time (it is not known if they paid him money or not). It was at that time the Bell Labs version was given the name bipolar junction transistor, or simply junction transistor, and Lilienfeld's design took the name field effect transistor.

In 1959, Dawon Kahng and Martin M. (John) Atalla at Bell Labs invented the metal–oxide–semiconductor field-effect transistor (MOSFET) as an offshoot to the patented FET design.Operationally and structurally different from the bipolar junction transistor, the MOSFET was made by putting an insulating layer on the surface of the semiconductor and then placing a metallic gate electrode on that. It used crystalline silicon for the semiconductor and a thermally oxidized layer of silicon dioxide for the insulator. The silicon MOSFET did not generate localized electron traps at the interface between the silicon and its native oxide layer, and thus was inherently free from the trapping and scattering of carriers that had impeded the performance of earlier field-effect transistors.

Discrete power MOSFETs are currently widely used as low voltage switches.

COMPOSITION

Usually the semiconductor of choice is silicon, but some chip manufacturers, most notably IBM and Intel, recently started using a chemical compound of silicon and germanium (SiGe) in MOSFET channels. Unfortunately, many semiconductors with better electrical properties than silicon, such as gallium arsenide, do not form good semiconductor-to-insulator interfaces, and thus are not suitable for MOSFETs. Research continues on creating insulators with acceptable electrical characteristics on other semiconductor material.

In order to overcome the increase in power consumption due to gate current leakage, a high-κ dielectric is used instead of silicon dioxide for the gate insulator, while polysilicon is replaced by metal gates.

The gate is separated from the channel by a thin insulating layer, traditionally of silicon dioxide and later of silicon oxynitride. Some companies have started to introduce a high-κ dielectric + metal gate combination in the 45 nanometer node.

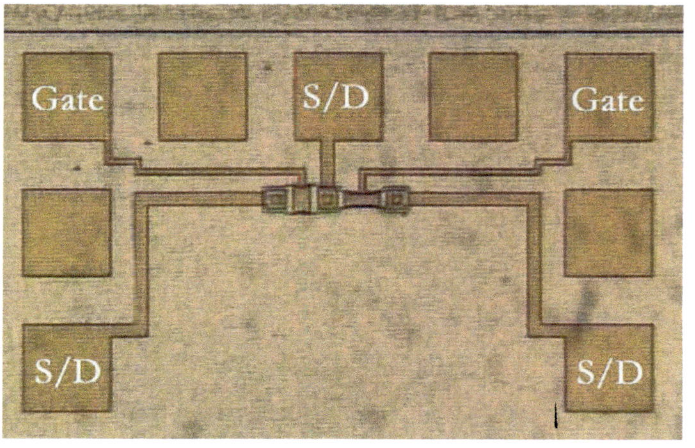

Photomicrograph of two metal-gate MOSFETs in a test pattern. Probe pads for two gates and three source/drain nodes are labeled.

When a voltage is applied between the gate and body terminals, the electric field generated penetrates through the oxide and creates an "inversion layer" or "channel" at the semiconductor-insulator interface. The inversion channel is of the same type, p-type or n-type, as the source and drain, and thus it provides a channel through which current can pass. Varying the voltage between the gate and body modulates the conductivity of this layer and thereby controls the current flow between drain and source. This is known as enhancement mode.

OPERATION

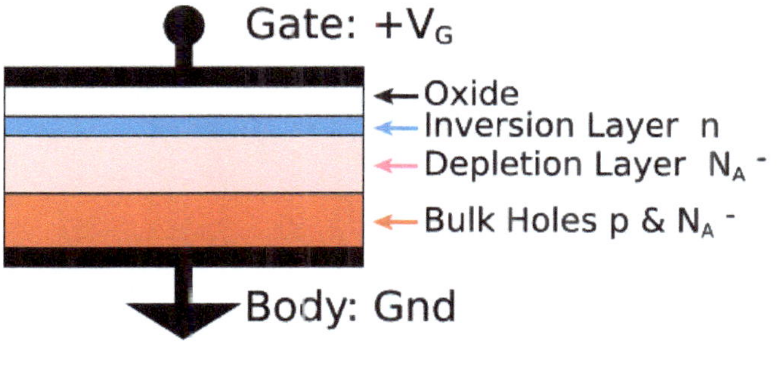

Metal–oxide–semiconductor structure on p-type silicon.

Metal–Oxide–Semiconductor Structure

The traditional metal–oxide–semiconductor (MOS) structure is obtained by growing a layer of silicon dioxide (SiO_2) on top of a silicon substrate and depositing a layer of metal or polycrystalline silicon (the latter is commonly used). As the silicon dioxide is a dielectric material, its structure is equivalent to a planar capacitor, with one of the electrodes replaced by a semiconductor.

When a voltage is applied across a MOS structure, it modifies the distribution of charges in the semiconductor. If we consider a p-type semiconductor (with N_A the density of acceptors, p the density of holes; $p = N_A$ in neutral bulk), a positive voltage, V_{GB}, from gate to body (see figure) creates a depletion layer by forcing the positively charged holes away from the gate-insulator/semiconductor interface, leaving exposed a carrier-free region of immobile, negatively charged acceptor ions. If V_{GB} is high enough, a high concentration of negative charge carriers forms in an *inversion layer* located in a thin layer next to the interface between the semiconductor and the insulator. Unlike the MOSFET, where the inversion layer electrons are supplied rapidly from the source/drain electrodes, in the MOS capacitor they are produced much more slowly by thermal generation through carrier generation and recombination centers in the depletion region. Conventionally, the gate voltage at which the volume density of electrons in the inversion layer is the same as the volume density of holes in the body is called the threshold voltage. When the voltage between transistor gate and source (V_{GS}) exceeds the threshold voltage (V_{th}), it is known as overdrive voltage.

This structure with p-type body is the basis of the n-type MOSFET, which requires the addition of n-type source and drain regions.

Structure and Channel Formation

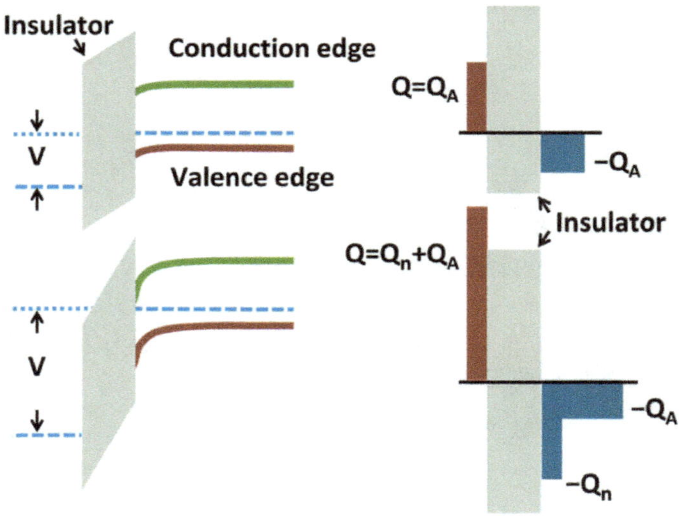

Channel formation in nMOS MOSFET shown as band diagram: Top panels: An applied gate voltage bends bands, depleting holes from surface (left). The charge inducing the bending is balanced by a layer of negative acceptor-ion charge (right). Bottom panel: A larger applied voltage further depletes holes but conduction band lowers enough in energy to populate a conducting channel

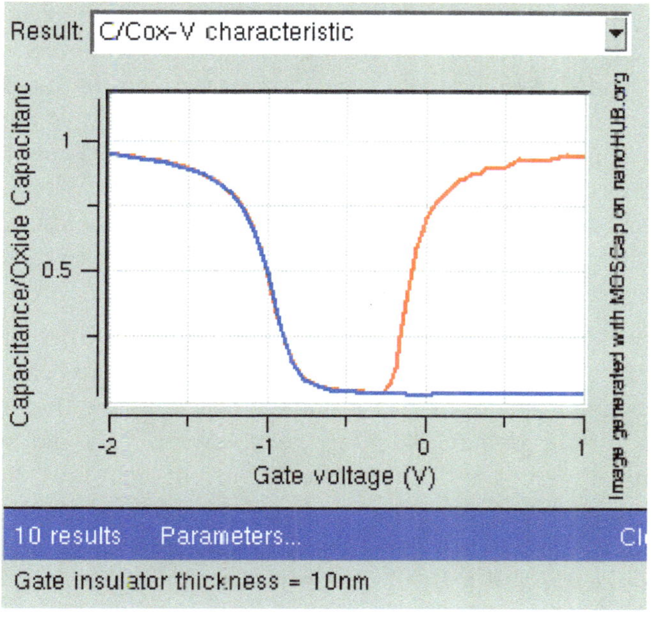

C–V profile for a bulk MOSFET with different oxide thickness. The leftmost part of the curve corresponds to accumulation. The valley in the middle corresponds to depletion. The curve on the right corresponds to inversion.

A MOSFET is based on the modulation of charge concentration by a MOS capacitance between a *body* electrode and a *gate* electrode located above the body and insulated from all other device regions by a gate dielectric layer which in the case of a MOSFET is an oxide, such as silicon dioxide. If dielectrics other than an oxide such as silicon dioxide (often referred to as oxide) are employed the device may be referred to as a metal–insulator–semiconductor FET (MISFET). Compared to the MOS capacitor, the MOSFET includes two additional terminals (*source* and *drain*), each connected to individual highly doped regions that are separated by the body region. These regions can be either p or n type, but they must both be of the same type, and of opposite type to the body region. The source and drain (unlike the body) are highly doped as signified by a "+" sign after the type of doping.

If the MOSFET is an n-channel or nMOS FET, then the source and drain are "n+" regions and the body is a "p" region. If the MOSFET is a p-channel or pMOS FET, then the source and drain are "p+" regions and the body is a "n" region. The source is so named because it is the source of the charge carriers (electrons for n-channel, holes for p-channel) that flow through the channel; similarly, the drain is where the charge carriers leave the channel.

The occupancy of the energy bands in a semiconductor is set by the position of the Fermi level relative to the semiconductor energy-band edges.

With sufficient gate voltage, the valence band edge is driven far from the Fermi level, and holes from the body are driven away from the gate.

At larger gate bias still, near the semiconductor surface the conduction band edge is brought close to the Fermi level, populating the surface with electrons in an *inversion layer* or *n-channel* at the interface between the p region and the oxide. This conducting channel extends between the source and the drain, and current is conducted through it when a voltage is applied between the two electrodes. Increasing the voltage on the gate leads to a higher electron density in the inversion layer and therefore increases the current flow between the source and drain. For gate voltages below the threshold value, the channel is lightly populated, and only a very small subthreshold leakage current can flow between the source and the drain.

When a negative gate-source voltage (positive source-gate) is applied, it creates a *p-channel* at the surface of the n region, analogous to the n-channel case, but with opposite polarities of charges and voltages. When a voltage less negative than the threshold value (a negative voltage for the p-channel) is applied between gate and source, the channel disappears and only a very small subthreshold current can flow between the source and the drain. The device may comprise a silicon on insulator (SOI) device in which a buried oxide (BOX) is formed below a thin semiconductor layer. If the channel region between the gate dielectric and a BOX region is very thin, the channel is referred to as an ultrathin channel (UTC) region with the source and drain regions formed on either side thereof in and/or above the thin semiconductor layer. Alternatively, the device may comprise a semiconductor on insulator (SEMOI) device in which semiconductors other than silicon are employed. Many alternative semiconductor materials may be employed. When the source and drain regions are formed above the channel in whole or in part, they are referred to as raised source/drain (RSD) regions.

CIRCUIT SYMBOLS

A variety of symbols are used for the MOSFET. The basic design is generally a line for the channel with the source and drain leaving it at right angles and then bending back at right angles into the same direction as the channel. Sometimes three line segments are used for enhancement mode and a solid line for depletion mode. Another line is drawn parallel to the channel for the gate.

The "bulk" or "body" connection, if shown, is shown connected to the back of the channel with an arrow indicating pMOS or nMOS. Arrows always point from P to N, so an NMOS (N-channel in P-well or P-substrate) has the arrow pointing in (from the bulk to the channel). If the bulk is connected to the source (as is generally the case with discrete devices) it is sometimes angled to meet up with the source leaving the transistor. If the bulk is not shown (as is often the case in IC design as they are generally common bulk) an inversion symbol is sometimes used to indicate PMOS, alternatively an arrow on the source may be used in the same way as for bipolar transistors (out for nMOS, in for pMOS).

Comparison of enhancement-mode and depletion-mode MOSFET symbols, along with JFET symbols. The orientation of the symbols, (most significantly the position of source relative to drain) is such that more positive voltages appear higher on the page than less positive voltages, implying current flowing "down" the page:

Comparison of n- and p-type MOSFETs		
Parameter	**nMOSFET**	**pMOSFET**
Source/drain type	n-type	p-type
Channel type (MOS capacitor)	n-type	p-type
Gate type (poly Si)	n+ poly-Si	p+ poly-Si
Gate type (metal)	φ_m ~ Si CB	φ_m ~ Si VB
Well type	p-type	n-type
Threshold voltage, V_{th}	positive (enhancement) negative (depletion)	negative (enhancement) positive (depletion)
Band-bending	Downwards	Upwards
Inversion layer carriers	electrons	holes
Substrate type	p-type	n-type

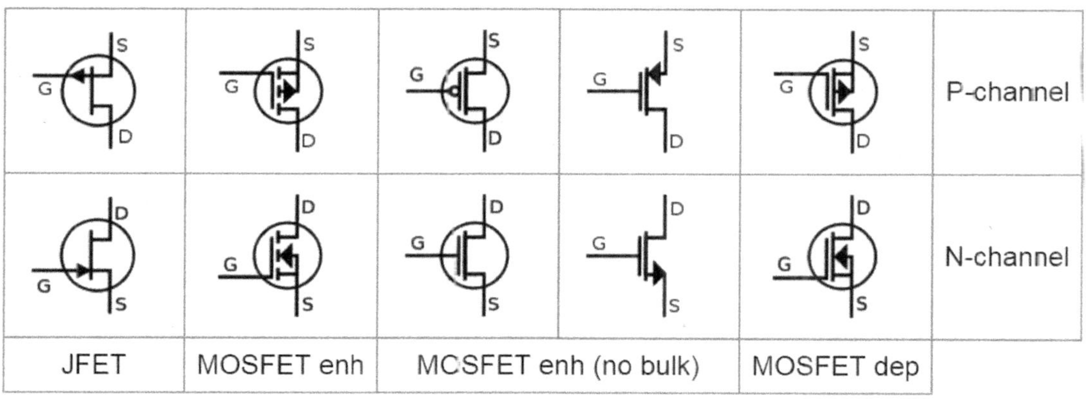

In schematics where G, S, D are not labeled, the detailed features of the symbol indicate which terminal is source and which is drain. For enhancement-mode and depletion-mode MOSFET symbols (in columns two and five), the source terminal is the one connected to the triangle. Additionally, in this diagram, the gate is shown as an "L" shape, whose input leg is closer to S than D, also indicating which is which. However, these symbols are often drawn with a "T" shaped gate (as elsewhere on this page), so it is the triangle which must be relied upon to indicate the source terminal.

For the symbols in which the bulk, or body, terminal is shown, it is here shown internally connected to the source (i.e., the black triangles in the diagrams in columns 2 and 5). This is a typical configuration, but by no means the only important configuration. In general, the MOSFET is a four-terminal device, and in integrated circuits many of the MOSFETs share a body connection, not necessarily connected to the source terminals of all the transistors.

APPLICATIONS

Digital integrated circuits such as microprocessors and memory devices contain thousands to millions of integrated MOSFET transistors on each device, providing the basic switching functions required to implement logic gates and data storage. Discrete devices are widely used in applications such as switch mode power supplies, variable-frequency drives and other power electronics applications where each device may be switching hundreds or thousands of watts. Radio-frequency amplifiers up to the UHF spectrum use MOSFET transistors as analog signal and power amplifiers. Radio systems also use MOSFETs as oscillators, or mixers to convert frequencies. MOSFET devices are also applied in audio-frequency power amplifiers for public address systems, sound reinforcement and home and automobile sound systems

MOS Integrated Circuits

Following the development of clean rooms to reduce contamination to levels never before thought necessary, and of photolithography and the planar process to allow circuits to be made in very few steps, the Si–SiO_2 system possessed such technical attractions as low cost of production (on a per circuit basis) and ease of integration. Largely because of these two factors, the MOSFET has become the most widely used type of transistor in integrated circuits.

General Microelectronics introduced the first commercial MOS integrated circuit in 1964.

Additionally, the method of coupling two complementary MOSFETS (P-channel and N-channel) into one high/low switch, known as CMOS, means that digital circuits dissipate very little power except when actually switched.

The earliest microprocessors starting in 1970 were all "MOS microprocessors"—i.e., fabricated entirely from PMOS logic or fabricated entirely from NMOS logic. In the 1970s, "MOS microprocessors" were often contrasted with "CMOS microprocessors" and "bipolar bit-slice processors".

CMOS Circuits

The MOSFET is used in digital complementary metal–oxide–semiconductor (CMOS) logic, which uses p- and n-channel MOSFETs as building blocks. Overheating is a major concern in integrated circuits since ever more transistors are packed into ever smaller chips. CMOS logic reduces power consumption because no current flows (ideally), and thus no power is consumed, except when the inputs to logic gates are being switched. CMOS accomplishes this current reduction by complementing every nMOSFET with a pMOSFET and connecting both gates and both drains together. A high voltage on the gates will cause the nMOSFET to conduct and the pMOSFET not to conduct and a low voltage on the gates causes the reverse. During the switching time as the voltage goes from one state to another, both MOSFETs will conduct briefly. This arrangement greatly reduces power consumption and heat generation.

Digital

The growth of digital technologies like the microprocessor has provided the motivation to advance MOSFET technology faster than any other type of silicon-based transistor. A big advantage of MOSFETs for digital switching is that the oxide layer between the gate and the channel prevents DC current from flowing through the gate, further reducing power consumption and giving a very large input impedance. The insulating oxide between the gate and channel effectively isolates a MOSFET in one logic stage from earlier and later stages, which allows a single MOSFET output to drive a considerable number of MOSFET inputs. Bipolar transistor-based logic (such as TTL) does not have such a high fanout capacity. This isolation also makes it easier for the designers to ignore to some extent loading effects between logic stages independently. That extent is defined by the operating frequency: as frequencies increase, the input impedance of the MOSFETs decreases.

Analog

The MOSFET's advantages in digital circuits do not translate into supremacy in all analog circuits. The two types of circuit draw upon different features of transistor behavior. Digital circuits switch, spending most of their time outside the switching region, while analog circuits depend on the linearity of response when the MOSFET is held precisely in the switching region. The bipolar junction transistor (BJT) has traditionally been the analog designer's transistor of choice, due largely to its higher transconductance and its lower output impedance (drain-voltage independence) in the switching region.

Nevertheless, MOSFETs are widely used in many types of analog circuits because of certain advantages. The characteristics and performance of many analog circuits can be scaled up or down by changing the sizes (length and width) of the MOSFETs used. By comparison, in most bipolar transistors the size of the device does not significantly affect its performance. MOSFETs' ideal characteristics regarding gate current (zero) and drain-source offset voltage (zero) also make them nearly ideal switch elements, and also make switched capacitor analog circuits practical. In their linear region, MOSFETs can be used as precision resistors, which can have a much higher controlled resistance than BJTs. In high power circuits, MOSFETs sometimes have the advantage of not suffering from thermal runaway as BJTs do. Also, MOSFETs can be configured to perform as capacitors and gyrator circuits which allow op-amps made from them to appear as inductors, thereby allowing all of the normal analog devices on a chip (except for diodes, which can be made smaller than a MOSFET anyway) to be built entirely out of MOSFETs. This means that complete analog circuits can be made on a silicon chip in a much smaller space and with simpler fabrication techniques. MOSFETS are ideally suited to switch inductive loads because of tolerance to inductive kickback.

Some ICs combine analog and digital MOSFET circuitry on a single mixed-signal integrated circuit, making the needed board space even smaller. This creates a need to isolate the analog circuits from the digital circuits on a chip level, leading to the use of isolation rings and Silicon-On-Insulator (SOI). Since MOSFETs require more space to handle a given amount of power than a BJT, fabrication processes can incorporate BJTs and MOSFETs into a single device. Mixed-transistor devices are called Bi-FETs (bipolar

FETs) if they contain just one BJT-FET and BiCMOS (bipolar-CMOS) if they contain complementary BJT-FETs. Such devices have the advantages of both insulated gates and higher current density.

Analog Switches

MOSFET analog switches use the MOSFET to pass analog signals when on, and as a high impedance when off. Signals flow in both directions across a MOSFET switch. In this application, the drain and source of a MOSFET exchange places depending on the relative voltages of the source/drain electrodes. The source is the more negative side for an N-MOS or the more positive side for a P-MOS. All of these switches are limited on what signals they can pass or stop by their gate–source, gate–drain and source–drain voltages; exceeding the voltage, current, or power limits will potentially damage the switch.

Single-Type

This analog switch uses a four-terminal simple MOSFET of either P or N type.

In the case of an n-type switch, the body is connected to the most negative supply (usually GND) and the gate is used as the switch control. Whenever the gate voltage exceeds the source voltage by at least a threshold voltage, the MOSFET conducts. The higher the voltage, the more the MOSFET can conduct. An N-MOS switch passes all voltages less than $V_{gate}-V_{tn}$. When the switch is conducting, it typically operates in the linear (or ohmic) mode of operation, since the source and drain voltages will typically be nearly equal.

In the case of a P-MOS, the body is connected to the most positive voltage, and the gate is brought to a lower potential to turn the switch on. The P-MOS switch passes all voltages higher than $V_{gate}-V_{tp}$ (threshold voltage V_{tp} is negative in the case of enhancement-mode P-MOS).

Dual-Type (CMOS)

This “complementary” or CMOS type of switch uses one P-MOS and one N-MOS FET to counteract the limitations of the single-type switch. The FETs have their drains and sources connected in parallel, the body of the P-MOS is connected to the high potential (V_{DD}) and the body of the N-MOS is connected to the low potential (Gnd). To turn the switch on, the gate of the P-MOS is driven to the low potential and the gate of the N-MOS is driven to the high potential. For voltages between $V_{DD}-V_{tn}$ and Gnd–V_{tp}, both FETs conduct the signal; for voltages less than Gnd–V_{tp}, the N-MOS conducts alone; and for voltages greater than $V_{DD}-V_{tn}$, the P-MOS conducts alone.

The voltage limits for this switch are the gate–source, gate–drain and source–drain voltage limits for both FETs. Also, the P-MOS is typically two to three times wider than the N-MOS, so the switch will be balanced for speed in the two directions.

Tri-state circuitry sometimes incorporates a CMOS MOSFET switch on its output to provide for a low-ohmic, full-range output when on, and a high-ohmic, mid-level signal when off.

CONSTRUCTION

Gate Material

The primary criterion for the gate material is that it is a good conductor. Highly doped polycrystalline silicon is an acceptable but certainly not ideal conductor, and also suffers from some more technical deficiencies in its role as the standard gate material. Nevertheless, there are several reasons favoring use of polysilicon:

- The threshold voltage (and consequently the drain to source on-current) is modified by the work function difference between the gate material and channel material. Because polysilicon is a semiconductor, its work function can be modulated by adjusting the type and level of doping. Furthermore, because polysilicon has the same bandgap as the underlying silicon channel, it is quite straightforward to tune the work function to achieve low threshold voltages for both NMOS and PMOS devices. By contrast, the work functions of metals are not easily modulated, so tuning the work function to obtain low threshold voltages becomes a significant challenge. Additionally, obtaining low-threshold devices on both PMOS and NMOS devices sometimes requires the use of different metals for each device type. While bimetallic integrated circuits (i.e., one type of metal for gate electrodes of NFETS and a second type of metal for gate electrodes of PFETS) are not common, they are known in patent literature and provide some benefit in terms of tuning electrical circuits' overall electrical perforamnce.
- The silicon-SiO_2 interface has been well studied and is known to have relatively few defects. By contrast many metal–insulator interfaces contain significant levels of defects which can lead to Fermi level pinning, charging, or other phenomena that ultimately degrade device performance.
- In the MOSFET IC fabrication process, it is preferable to deposit the gate material prior to certain high-temperature steps in order to make better-performing transistors. Such high temperature steps would melt some metals, limiting the types of metal that can be used in a metal-gate-based process.

While polysilicon gates have been the de facto standard for the last twenty years, they do have some disadvantages which have led to their likely future replacement by metal gates. These disadvantages include:

- Polysilicon is not a great conductor (approximately 1000 times more resistive than metals) which reduces the signal propagation speed through the material. The resistivity can be lowered by increasing the level of doping, but even highly doped polysilicon is not as conductive as most metals. To improve conductivity further, sometimes a high-temperature metal such as tungsten, titanium, cobalt, and more recently nickel is alloyed with the top layers of the polysilicon. Such a blended material is called silicide. The silicide-polysilicon combination has better electrical properties than polysilicon alone and still does not melt in subsequent processing. Also the threshold voltage is not significantly higher than with polysilicon alone, because the silicide material is not near the channel. The process in which silicide is formed on both the gate electrode and the source and drain regions is sometimes called salicide, self-aligned silicide.

- When the transistors are extremely scaled down, it is necessary to make the gate dielectric layer very thin, around 1 nm in state-of-the-art technologies. A phenomenon observed here is the so-called poly depletion, where a depletion layer is formed in the gate polysilicon layer next to the gate dielectric when the transistor is in the inversion. To avoid this problem, a metal gate is desired. A variety of metal gates such as tantalum, tungsten, tantalum nitride, and titanium nitride are used, usually in conjunction with high-k dielectrics. An alternative is to use fully silicided polysilicon gates, a process known as FUSI.

Present high performance CPUs use metal gate technology, together with high-k dielectrics, a combination known as HKMG (High-K, Metal Gate). The disadvantages of metal gates are overcome by a few techniques:

- The threshold voltage is tuned by including a thin "work function metal" layer between the high-K dielectric and the main metal. This layer is thin enough that the total work function of the gate is influenced by both the main metal and thin metal work functions (either due to alloying during annealing, or simply due to the incomplete screening by the thin metal). The threshold voltage thus can be tuned by the thickness of the thin metal layer.
- High-K dielectrics are now well studied, and their defects are understood.
- HKMG processes exist that do not require the metals to experience high temperature anneals; other processes select metals that can survive the annealing step.

Insulator

As devices are made smaller, insulating layers are made thinner, and at some point tunneling of carriers through the insulator from the channel to the gate electrode takes place. To reduce the resulting leakage current, the insulator can be made thicker by choosing a material with a higher dielectric constant. To see how thickness and dielectric constant are related, note that Gauss's law connects field to charge as:

$$Q = \kappa \epsilon_0 \, E,$$

with Q = charge density, κ = dielectric constant, ε_0 = permittivity of empty space and E = electric field. From this law it appears the same charge can be maintained in the channel at a lower field provided κ is increased. The voltage on the gate is given by:

$$V_G = V_{ch} + E \, t_{ins} = V_{ch} + \frac{Q t_{ins}}{\kappa \epsilon_0},$$

with V_G = gate voltage, V_{ch} = voltage at channel side of insulator, and t_{ins} = insulator thickness. This equation shows the gate voltage will not increase when the insulator thickness increases, provided κ increases to keep $t_{ins}/\kappa = constant$.

The insulator in a MOSFET is a dielectric which can in any event be silicon oxide, but many other dielectric materials are employed. The generic term for the dielectric is gate dielectric since the dielectric lies directly below the gate electrode and above the channel of the MOSFET.

Junction Design

The source-to-body and drain-to-body junctions are the object of much attention because of three major factors: their design affects the current-voltage (*I-V*) characteristics of the device, lowering output resistance, and also the speed of the device through the loading effect of the junction capacitances, and finally, the component of stand-by power dissipation due to junction leakage.

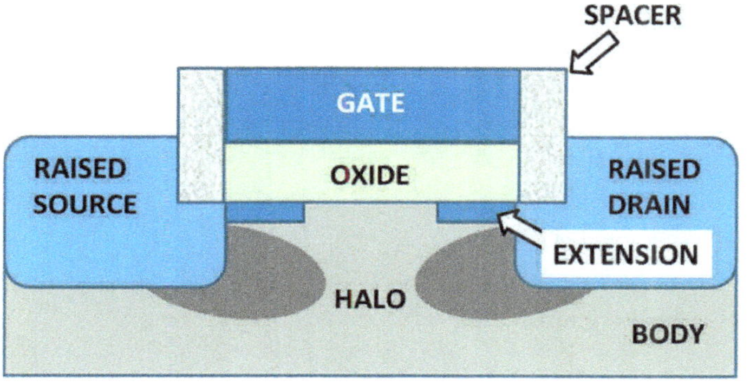

MOSFET showing shallow junction extensions, raised source and drain and halo implant. Raised source and drain separated from gate by oxide spacers

The drain induced barrier lowering of the threshold voltage and channel length modulation effects upon *I-V* curves are reduced by using shallow junction extensions. In addition, *halo* doping can be used, that is, the addition of very thin heavily doped regions of the same doping type as the body tight against the junction walls to limit the extent of depletion regions.

The capacitive effects are limited by using raised source and drain geometries that make most of the contact area border thick dielectric instead of silicon.

These various features of junction design are shown (with artistic license) in the figure.

Scaling

Over the past decades, the MOSFET has continually been scaled down in size; typical MOSFET channel lengths were once several micrometres, but modern integrated circuits are incorporating MOSFETs with channel lengths of tens of nanometers. Robert Dennard's work on scaling theory was pivotal in recognizing that this ongoing reduction was possible. Intel began production of a process featuring a 32 nm feature size (with the channel being even shorter) in late 2009. The semiconductor industry maintains a "roadmap", the ITRS, which sets the pace for MOSFET development. Historically, the difficulties with decreasing the size of the MOSFET have been associated with the semiconductor device fabrication process, the need to use very low voltages, and with poorer electrical performance necessitating circuit redesign and innovation (small MOSFETs exhibit higher leakage currents and lower output resistance).

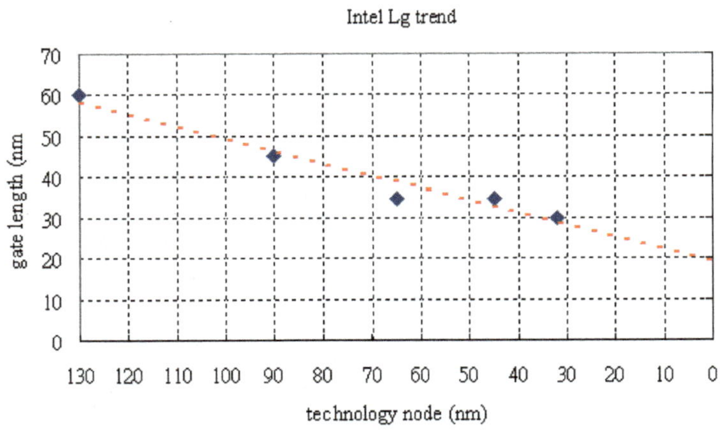

Trend of Intel CPU transistor gate length

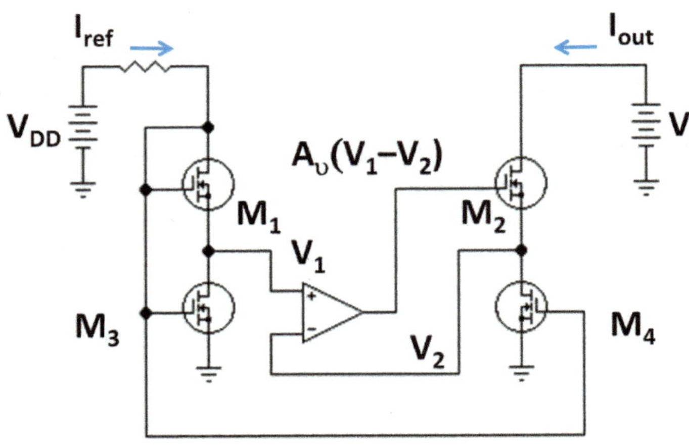

MOSFET version of gain-boosted current mirror; M_1 and M_2 are in active mode, while M_3 and M_4 are in Ohmic mode, and act like resistors. The operational amplifier provides feedback that maintains a high output resistance

Smaller MOSFETs are desirable for several reasons. The main reason to make transistors smaller is to pack more and more devices in a given chip area. This results in a chip with the same functionality in a smaller area, or chips with more functionality in the same area. Since fabrication costs for a semiconductor wafer are relatively fixed, the cost per integrated circuits is mainly related to the number of chips that can be produced per wafer. Hence, smaller ICs allow more chips per wafer, reducing the price per chip. In fact, over the past 30 years the number of transistors per chip has been doubled every 2–3 years once a new technology node is introduced. For example, the number of MOSFETs in a microprocessor fabricated in a 45 nm technology can well be twice as many as in a 65 nm chip. This doubling of transistor density was first observed by Gordon Moore in 1965 and is commonly referred to as Moore's law. It is also expected that smaller transistors switch faster. For example, one approach to size reduction is a scaling of the MOSFET that requires all device dimensions to reduce proportionally. The main device dimensions are the channel length, channel width, and oxide thickness. When they are scaled down by equal factors, the transistor channel resistance does not change, while gate capacitance is cut by that factor. Hence, the RC delay of the transistor scales with a similar factor. While this has been traditionally the case for the older technologies, for the state-of-the-art MOSFETs reduction of the transistor dimensions does not necessarily translate to higher chip speed because the delay due to interconnections is more significant.

Producing MOSFETs with channel lengths much smaller than a micrometre is a challenge, and the difficulties of semiconductor device fabrication are always a limiting factor in advancing integrated circuit technology. Though processes such as ALD have improved fabrication for small components, the small size of the MOSFET (less than a few tens of nanometers) has created operational problems:

Higher Subthreshold Conduction

As MOSFET geometries shrink, the voltage that can be applied to the gate must be reduced to maintain reliability. To maintain performance, the threshold voltage of the MOSFET has to be reduced as well. As threshold voltage is reduced, the transistor cannot be switched from complete turn-off to complete turn-on with the limited voltage swing available; the circuit design is a compromise between strong current in the "on" case and low current in the "off" case, and the application determines whether to favor one over the other. Subthreshold leakage (including subthreshold conduction, gate-oxide leakage and reverse-biased junction leakage), which was ignored in the past, now can consume upwards of half of the total power consumption of modern high-performance VLSI chips.

Increased Gate-Oxide Leakage

The gate oxide, which serves as insulator between the gate and channel, should be made as thin as possible to increase the channel conductivity and performance when the transistor is on and to reduce subthreshold leakage when the transistor is off. However, with current gate oxides with a thickness of around 1.2 nm (which in silicon is ~5 atoms thick) the quantum mechanical phenomenon of electron tunneling occurs between the gate and channel, leading to increased power consumption. Silicon dioxide has traditionally been used as the gate insulator. Silicon dioxide however has a modest dielectric constant. Increasing the dielectric constant of the gate dielectric allows a thicker layer while maintaining a high capacitance (capacitance is proportional to dielectric constant and inversely proportional to dielectric thickness). All else equal, a higher dielectric thickness reduces the quantum tunneling current through the dielectric between the gate and the channel. Insulators that have a larger dielectric constant than silicon dioxide (referred to as high-k dielectrics), such as group IVb metal silicates e.g. hafnium and zirconium silicates and oxides are being used to reduce the gate leakage from the 45 nanometer technology node onwards. On the other hand, the barrier height of the new gate insulator is an important consideration; the difference in conduction band energy between the semiconductor and the dielectric (and the corresponding difference in valence band energy) also affects leakage current level. For the traditional gate oxide, silicon dioxide, the former barrier is approximately 8 eV. For many alternative dielectrics the value is significantly lower, tending to increase the tunneling current, somewhat negating the advantage of higher dielectric constant. The maximum gate-source voltage is determined by the strength of the electric field able to be sustained by the gate dielectric before significant leakage occurs. As the insulating dielectric is made thinner, the electric field strength within it goes up for a fixed voltage. This necessitates using lower voltages with the thinner dielectric.

Increased Junction Leakage

To make devices smaller, junction design has become more complex, leading to higher doping levels, shallower junctions, "halo" doping and so forth, all to decrease drain-induced barrier lowering. To keep these complex junctions in place, the annealing steps formerly used to remove damage and electrically active defects must be curtailed increasing junction leakage. Heavier doping is also associated with thinner depletion layers and more recombination centers that result in increased leakage current, even without lattice damage.

DIBL and V_T Roll Off

Because of the short-channel effect, channel formation is not entirely done by the gate, but now the drain and source also affect the channel formation. As the channel length decreases, the depletion regions of the source and drain come closer together and make the threshold voltage (V_T) a function of the length of the channel. This is called V_T roll-off. V_T also becomes function of drain to source voltage V_{DS}. As we increase the V_{DS}, the depletion regions increase in size, and a considerable amount of charge is depleted by the V_{DS}. The gate voltage required to form the channel is then lowered, and thus, the V_T decreases with an increase in V_{DS}. This effect is called drain induced barrier lowering (DIBL).

Lower Output Resistance

For analog operation, good gain requires a high MOSFET output impedance, which is to say, the MOSFET current should vary only slightly with the applied drain-to-source voltage. As devices are made smaller, the influence of the drain competes more successfully with that of the gate due to the growing proximity of these two electrodes, increasing the sensitivity of the MOSFET current to the drain voltage. To counteract the resulting decrease in output resistance, circuits are made more complex, either by requiring more devices, for example the cascode and cascade amplifiers, or by feedback circuitry using operational amplifiers, for example a circuit like that in the adjacent figure.

Lower Transconductance

The transconductance of the MOSFET decides its gain and is proportional to hole or electron mobility (depending on device type), at least for low drain voltages. As MOSFET size is reduced, the fields in the channel increase and the dopant impurity levels increase. Both changes reduce the carrier mobility, and hence the transconductance. As channel lengths are reduced without proportional reduction in drain voltage, raising the electric field in the channel, the result is velocity saturation of the carriers, limiting the current and the transconductance.

Interconnect Capacitance

Traditionally, switching time was roughly proportional to the gate capacitance of gates. However, with transistors becoming smaller and more transistors being placed on the chip, interconnect capacitance (the capacitance of the metal-layer connections between different parts of the chip) is becoming a large percentage of capacitance. Signals have to travel through the interconnect, which leads to increased delay and lower performance.

Heat Production

The ever-increasing density of MOSFETs on an integrated circuit creates problems of substantial localized heat generation that can impair circuit operation. Circuits operate more slowly at high temperatures, and have reduced reliability and shorter lifetimes. Heat sinks and other cooling devices and methods are now required for many integrated circuits including microprocessors. Power MOSFETs are at risk of thermal runaway. As their on-state resistance rises with temperature, if the load is approximately a constant-current load then the power loss rises correspondingly, generating further heat. When the heatsink is not able to keep the temperature low enough, the junction temperature may rise quickly and uncontrollably, resulting in destruction of the device.

Process Variations

With MOSFETs becoming smaller, the number of atoms in the silicon that produce many of the transistor's properties is becoming fewer, with the result that control of dopant numbers and placement is more erratic. During chip manufacturing, random process variations affect all transistor dimensions: length, width, junction depths, oxide thickness *etc.*, and become a greater percentage of overall transistor size as the transistor shrinks. The transistor characteristics become less certain, more statistical. The random nature of manufacture means we do not know which particular example MOSFETs actually will end up in a particular instance of the circuit. This uncertainty forces a less optimal design because the design must work for a great variety of possible component MOSFETs.

Modeling Challenges

Modern ICs are computer-simulated with the goal of obtaining working circuits from the very first manufactured lot. As devices are miniaturized, the complexity of the processing makes it difficult to predict exactly what the final devices look like, and modeling of physical processes becomes more challenging as well. In addition, microscopic variations in structure due simply to the probabilistic nature of atomic processes require statistical (not just deterministic) predictions. These factors combine to make adequate simulation and "right the first time" manufacture difficult.

Other Types

Dual-Gate

The dual-gate MOSFET has a tetrode configuration, where both gates control the current in the device. It is commonly used for small-signal devices in radio frequency applications where biasing the drain-side gate at constant potential reduces the gain loss caused by Miller effect, replacing two separate transistors in cascode configuration. Other common uses in RF circuits include gain control and mixing (frequency conversion). The "tetrode" description, though accurate, does not replicate the vacuum-tube tetrode. Vacuum-tube tetrodes, using a screen grid, exhibit much lower grid-plate capacitance and much higher output impedance and voltage gains than triode vacuum tubes. These improvements are commonly an order of magnitude (10 times) or considerably more. Tetrode transistors (whether bipolar junction or field-effect) do not exhibit improvements of such a great degree.

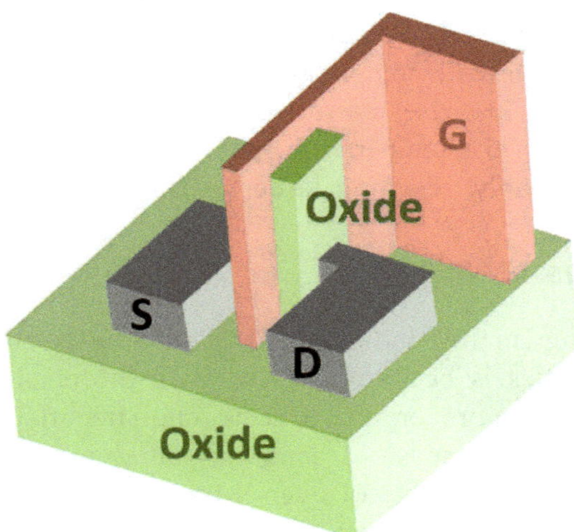

A FinFET

The FinFET is a double-gate silicon-on-insulator device, one of a number of geometries being introduced to mitigate the effects of short channels and reduce drain-induced barrier lowering. The "fin" refers to the narrow channel between source and drain. A thin insulating oxide layer on either side of the fin separates it from the gate. SOI FinFETs with a thick oxide on top of the fin are called *double-gate* and those with a thin oxide on top as well as on the sides are called *triple-gate* FinFETs.

Depletion-Mode

There are *depletion-mode* MOSFET devices, which are less commonly used than the standard *enhancement-mode* devices already described. These are MOSFET devices that are doped so that a channel exists even with zero voltage from gate to source. To control the channel, a negative voltage is applied to the gate (for an n-channel device), depleting the channel, which reduces the current flow through the device. In essence, the depletion-mode device is equivalent to a normally closed (on) switch, while the enhancement-mode device is equivalent to a normally open (off) switch. Due to their low noise figure in the RF region, and better gain, these devices are often preferred to bipolars in RF front-ends such as in TV sets. Depletion-mode MOSFET families include BF 960 by Siemens and BF 980 by Philips (dated 1980s), whose derivatives are still used in AGC and RF mixer front-ends.

NMOS Logic

For devices of equal current driving capability, n-channel MOSFETs can be made smaller than p-channel MOSFETs, due to p-channel charge carriers (holes) having lower mobility than do n-channel charge carriers (electrons), and producing only one type of MOSFET on a silicon substrate is cheaper and technically simpler. These were the driving principles in the design of NMOS logic which uses n-channel MOSFETs exclusively. However, neglecting leakage current, unlike CMOS logic, NMOS logic consumes power even when no switching is taking place. With advances in technology, CMOS logic displaced NMOS logic in the mid-1980s to become the preferred process for digital chips.

Power MOSFET

Power MOSFETs have a different structure. As with most power devices, the structure is vertical and not planar. Using a vertical structure, it is possible for the transistor to sustain both high blocking voltage and high current. The voltage rating of the transistor is a function of the doping and thickness of the N-epitaxial layer (see cross section), while the current rating is a function of the channel width (the wider the channel, the higher the current). In a planar structure, the current and breakdown voltage ratings are both a function of the channel dimensions (respectively width and length of the channel), resulting in inefficient use of the "silicon estate". With the vertical structure, the component area is roughly proportional to the current it can sustain, and the component thickness (actually the N-epitaxial layer thickness) is proportional to the breakdown voltage.

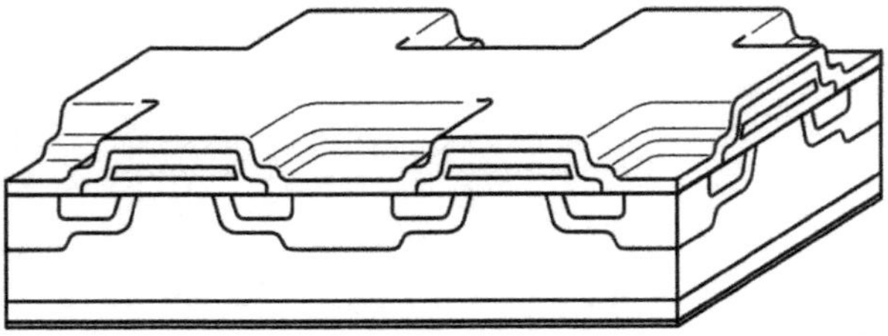

Cross section of a power MOSFET, with square cells. A typical transistor is constituted of several thousand cells

Power MOSFETs with lateral structure are mainly used in high-end audio amplifiers and high-power PA systems. Their advantage is a better behaviour in the saturated region (corresponding to the linear region of a bipolar transistor) than the vertical MOSFETs. Vertical MOSFETs are designed for switching applications.

DMOS

DMOS stands for double-diffused metal–oxide–semiconductor. There are *LDMOS* (Lateral Double-diffused Metal Oxide Semiconductor) and *VDMOS* (Vertical Double-diffused Metal Oxide Semiconductor). Most power MOSFETs are made using this technology.

RHBD

Semiconductor sub-micrometer and nanometer electronic circuits are the primary concern for operating within the normal tolerance in harsh radiation environments like outer space. One of the design approaches for making a radiation-hardened-by-design (RHBD) device is Enclosed-Layout-Transistor (ELT). Normally, the gate of the MOSFET surrounds the drain, which is placed in the center of the ELT. The source of the MOSFET surrounds the gate. Another RHBD MOSFET is called H-Gate. Both of these transistors have very low leakage current with respect to radiation. However, they are large in size and take more space on silicon than a standard MOSFET. In older STI (shallow trench isolation) designs, radiation strikes near the silicon oxide region cause the channel inversion at the corners of the standard MOSFET due to accumulation of radiation induced trapped charges. If the charges are large enough, the accumulated charges affect STI surface edges along the channel near the channel interface (gate) of the standard MOSFET. Thus the device channel inversion occurs along the channel edges and the device creates off-state leakage path, causing device to turn on. So the reliability of circuits degrades severely. The ELT offers many advantages. These advantages include improvement of reliability by reducing unwanted surface inversion at the gate edges that occurs in the standard MOSFET. Since the gate edges are enclosed in ELT, there is no gate oxide edge (STI at gate interface), and thus the transistor off-state leakage is reduced very much. Low-power microelectronic circuits including computers, communication devices and monitoring systems in space shuttle and satellites are very different from what we use on earth. They are radiation (high-speed atomic particles like proton and neutron, solar flare magnetic energy dissipation in Earth's space, energetic cosmic rays like X-ray, gamma ray etc.) tolerant circuits. These special electronics are designed by applying very different techniques using RHBD MOSFETs to ensure the safe space journey and also space-walk of astronauts.

OPERATIONAL AMPLIFIER

An operational amplifier (often op-amp or opamp) is a DC-coupled high-gain electronic voltage amplifier with a differential input and, usually, a single-ended output. In this configuration, an op-amp produces an output potential (relative to circuit ground) that is typically hundreds of thousands of times larger than the potential difference between its input terminals. Operational amplifiers had their origins in analog computers, where they were used to perform mathematical operations in many linear, non-linear and frequency-dependent circuits. The popularity of the op-amp as a building block in analog circuits is due to its versatility. Due to negative feedback, the characteristics of an op-amp circuit, its gain, input and output impedance, bandwidth etc. are determined by external components and have little dependence on temperature coefficients or manufacturing variations in the op-amp itself.

Op-amps are among the most widely used electronic devices today, being used in a vast array of consumer, industrial, and scientific devices. Many standard IC op-amps cost only a few cents in moderate production volume; however some integrated or hybrid operational amplifiers with special performance specifications may cost over $100 US in small quantities. Op-amps may be packaged as components, or used as elements of more complex integrated circuits.

The op-amp is one type of differential amplifier. Other types of differential amplifier include the fully differential amplifier (similar to the op-amp, but with two outputs), the instrumentation amplifier (usually built from three op-amps), the isolation amplifier (similar to the instrumentation amplifier, but with tolerance to common-mode voltages that would destroy an ordinary op-amp), and negative feedback amplifier (usually built from one or more op-amps and a resistive feedback network).

Internal Circuitry of 741-type op-amp

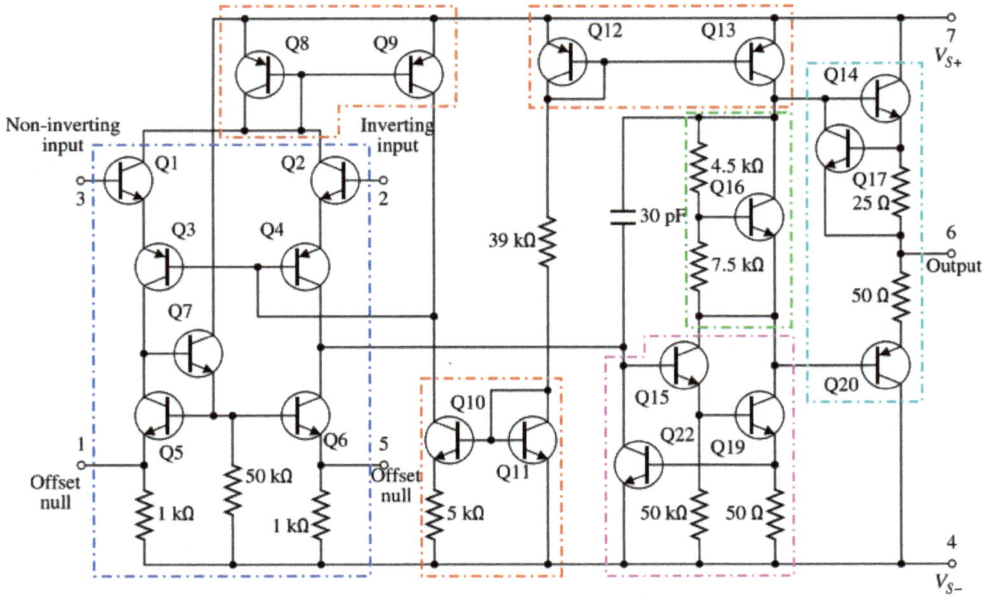

A component-level diagram of the common 741 op-amp. Dotted lines outline: current mirrors (red); differential amplifier (blue); class A gain stage (magenta); voltage level shifter (green); output stage (cyan).

Sourced by many manufacturers, and in multiple similar products, an example of a bipolar transistor operational amplifier is the 741 integrated circuit designed in 1968 by David Fullagar at Fairchild Semiconductor after Bob Widlar's LM301 integrated circuit design. In this discussion, we use the parameters of the Hybrid-pi model to characterize the small-signal, grounded emitter characteristics of a transistor. In this model, the current gain of a transistor is denoted h_{fe}, more commonly called the β.

Architecture

A small-scale integrated circuit, the 741 op-amp shares with most op-amps an internal structure consisting of three gain stages:

- Differential amplifier (outlined blue) — provides high differential amplification (gain), with rejection of common-mode signal, low noise, high input impedance, and drives a
- Voltage amplifier (outlined magenta) — provides high voltage gain, a single-pole frequency roll-off, and in turn drives the
- Output amplifier (outlined cyan and green) — provides high current gain (low output impedance), along with output current limiting, and output short-circuit protection.

Additionally, it contains current mirror (outlined red) bias circuitry and compensation capacitor (30 pF).

Differential Amplifier

The input stage consists of a cascaded differential amplifier (outlined in blue) followed by a current-mirror active load. This constitutes a transconductance amplifier, turning a differential voltage signal at the bases of Q1, Q2 into a current signal into the base of Q15.

It entails two cascaded transistor pairs, satisfying conflicting requirements. The first stage consists of the matched NPN emitter follower pair Q1, Q2 that provide high input impedance. The second is the matched PNP common-base pair Q3, Q4 that eliminates the undesirable Miller effect; it drives an active load Q7 plus matched pair Q5, Q6.

That active load is implemented as a modified Wilson current mirror; its role is to convert the (differential) input current signal to a single-ended signal without the attendant 50% losses (increasing the op-amp's open-loop gain by 3 dB). Thus, a small-signal differential current in Q3 versus Q4 appears summed (doubled) at the base of Q15, the input of the voltage gain stage.

Voltage Amplifier

The (class-A) voltage gain stage (outlined in magenta) consists of the two NPN transistors Q15/Q19 connected in a Darlington configuration and uses the output side of current mirror Q12/Q13 as its collector (dynamic) load to achieve its high voltage gain. The output sink transistor Q20 receives its base drive from the common collectors of Q15 and Q19; the level-shifter Q16 provides base drive for the output source transistor Q14. .

The transistor Q22 prevents this stage from delivering excessive current to Q20 and thus limits the output sink current.

Output Amplifier

The output stage (Q14, Q20, outlined in cyan) is a Class AB push-pull emitter follower amplifier. It provides an output drive with impedance of $\approx 50\Omega$, in essence, current gain. Transistor Q16 (outlined in green) provides the quiescent current for the output transistors, and Q17 provides output current limiting.

Biasing Circuits

Provide appropriate quiescent current for each stage of the op-amp.

The resistor (39 kΩ) connecting the (diode-connected) Q11 and Q12, and the given supply voltage ($V_{S+}-V_{S-}$), determine the current in the current mirrors, (matched pairs) Q10/Q11 and Q12/Q13. The collector current of Q11, i_{11} * 39 kΩ = $V_{S+} - V_{S-} - 2\ V_{BE}$. For the typical $V_S = \pm 20$ V, the standing current in Q11/Q12 (as well as in Q13) would be ≈ 1 mA. A supply current for a typical 741 of about 2 mA agrees with the notion that these two bias currents dominate the quiescent supply current.

Transistors Q11 and Q10 form a Widlar current mirror, with quiescent current in Q10 i_{10} such that $\ln(i_{11} / i_{10}) = i_{10} * 5\ \text{k}\Omega / 28\ \text{mV}$, where 5 kΩ represents the emitter resistor of Q10, and 28 mV is V_T, the thermal voltage at room temperature. In this case $i_{10} \approx 20\ \mu\text{A}$.

Differential Amplifier

The biasing circuit of this stage is set by a feedback loop that forces the collector currents of Q10 and Q9 to (nearly) match. The small difference in these currents provides the drive for the common base of Q3/Q4 (note that the base drive for input transistors Q1/Q2 is the input bias current and must be sourced externally). The summed quiescent currents of Q1/Q3 plus Q2/Q4 is mirrored from Q8 into Q9, where it is summed with the collector current in Q10, the result being applied to the bases of Q3/Q4.

The quiescent currents of Q1/Q3 (resp., Q2/Q4) i_1 will thus be half of i_{10}, of order $\approx$ 10 μA. Input bias current for the base of Q1 (resp. Q2) will amount to i_1 / β; typically $\approx$50 nA, implying a current gain $h_{fe} \approx 200$ for Q1(Q2).

This feedback circuit tends to draw the common base node of Q3/Q4 to a voltage $V_{com} - 2 * V_{BE}$, where V_{com} is the input common-mode voltage. At the same time, the magnitude of the quiescent current is relatively insensitive to the characteristics of the components Q1–Q4, such as h_{fe}, that would otherwise cause temperature dependence or part-to-part variations.

Transistor Q7 drives Q5 and Q6 into conduction until their (equal) collector currents match that of Q1/Q3 and Q2/Q4. The quiescent current in Q7 is V_{BE} / 50 kΩ, about 35μA, as is the quiescent current in Q15, with its matching operating point. Thus, the quiescent currents are pairwise matched in Q1/Q2, Q3/Q4, Q5/Q6, and Q7/Q15.

Voltage Amplifier

Quiescent currents in Q16 and Q19 are set by the current mirror Q12/Q13, which is running at $\approx$ 1 mA. Through some] mechanism, the collector current in Q19 tracks that standing current.

Output Amplifier

In the circuit involving Q16 (variously named rubber diode or V_{BE} multiplier), the 4.5 kΩ resistor must be conducting about 100 μA, with the Q16 V_{BE} roughly 700 mV. Then the V_{CB} must be about 0.45 V and V_{CE} at about 1.0 V. Because the Q16 collector is driven by a current source and the Q16 emitter drives into the Q19 collector current sink, the Q16 transistor establishes a voltage difference between Q14 base and Q20 base of $\approx$ 1 V, regardless of the common-mode voltage of Q14/Q20 base. The standing current in Q14/Q20 will be a factor $\exp(100\ \text{mV} / V_T) \approx 36$ smaller than the 1 mA quiescent current in the class A portion of the op amp. This (small) standing current in the output transistors establishes the output stage in class AB operation and reduces the crossover distortion of this stage.

Small-Signal Differential Mode

A small differential input voltage signal gives rise, through multiple stages of current amplification, to a much larger voltage signal on output.

Input Impedance

The input stage with Q1 and Q3 is similar to an emitter-coupled pair (long-tailed pair), with Q2 and Q4 adding some degenerating impedance. The input impedance is relatively high because of the small current through Q1-Q4. A typical 741 op amp has an differential input impedance of about 2 MΩ. The common mode input impedance is even higher, as the input stage works at an essentially constant current.

Differential Amplifier

A differential voltage V_{In} at the op-amp inputs (pins 3 and 2, respectively) gives rise to a small differential current in the bases of Q1 and Q2 $i_{In} \approx V_{In}$ / (2 h_{ie} * h_{fe}). This differential base current causes a change in the differential collector current in each leg by i_{In} * h_{fe}. Introducing the transconductance of Q1, $g_m = h_{fe} / h_{ie}$, the (small-signal) current at the base of Q15 (the input of the voltage gain stage) is V_{In} * g_m / 2.

This portion of the op amp cleverly changes a differential signal at the op amp inputs to a single-ended signal at the base of Q15, and in a way that avoids wastefully discarding the signal in either leg. To see how, notice that a small negative change in voltage at the inverting input (Q2 base) drives it out of conduction, and this incremental decrease in current passes directly from Q4 collector to its emitter, resulting in an decrease in base drive for Q15. On the other hand, a small positive change in voltage at the non-inverting input (Q1 base) drives this transistor into conduction, reflected in an increase in current at the collector of Q3. This current drives Q7 further into conduction, which turns on current mirror Q5/Q6. Thus, the increase in Q3 emitter current is mirrored in an increase in Q6 collector current; the increased collector currents shunts more from the collector node and results in a decrease in base drive current for Q15. Besides avoiding wasting 3 dB of gain here, this technique decreases common-mode gain and feedthrough of power supply noise.

Voltage Amplifier

A current signal i at Q15's base gives rise to a current in Q19 of order i * β^2 (the product of the h_{fe} of each of Q15 and Q19, which are connected in a Darlington pair). This current signal develops a voltage at the bases of output transistors Q14/Q20 proportional to the h_{ie} of the respective transistor.

Output Amplifier

Output transistors Q14 and Q20 are each configured as an emitter follower, so no voltage gain occurs there; instead, this stage provides current gain, equal to the h_{fe} of Q14 (resp. Q20).

The output impedance is not zero, as it would be in an ideal op-amp, but with negative feedback it approaches zero at low frequencies.

Overall Open-Loop Voltage Gain

The net open-loop small-signal voltage gain of the op amp involves the product of the current gain h_{fe} of some 4 transistors. In practice, the voltage gain for a typical 741-style op amp is of order 200,000, and the current gain, the ratio of input impedance (≈2−6 MΩ) to output impedance (≈50Ω) provides yet more (power) gain.

Other Linear Characteristics

Small-Signal Common Mode Gain

The ideal op amp has infinite common-mode rejection ratio, or zero common-mode gain.

In the present circuit, if the input voltages change in the same direction, the negative feedback makes Q3/Q4 base voltage follow (with $2V_{BE}$ below) the input voltage variations. Now the output part (Q10) of Q10-Q11 current mirror keeps up the common current through Q9/Q8 constant in spite of varying voltage. Q3/Q4 collector currents, and accordingly the output current at the base of Q15, remain unchanged.

In the typical 741 op amp, the common-mode rejection ratio is 90 dB, implying an open-loop common-mode voltage gain of about 6.

Frequency Compensation

The innovation of the Fairchild μA741 was the introduction of frequency compensation via an on-chip (monolithic) capacitor, simplifying application of the op amp by eliminating the need for external components for this function. The 30 pF capacitor stabilizes the amplifier via Miller compensation and functions in a manner similar to an op-amp integrator circuit. Also known as ‹dominant pole compensation› because it introduces a pole that masks (dominates) the effects of other poles into the open loop frequency response; in a 741 op amp this pole can be as low as 10 Hz (where it causes a −3 dB loss of open loop voltage gain).

This internal compensation is provided to achieve unconditional stability of the amplifier in negative feedback configurations where the feedback network is non-reactive and the closed loop gain is unity or higher. By contrast, amplifiers requiring external compensation, such as the μA748, may require external compensation or closed-loop gains significantly higher than unity.

Input Offset Voltage

The "offset null" pins may be used to place external resistors (typically in the form of the two ends of a potentiometer, with the slider connected to V_{S-}) in parallel with the emitter resistors of Q5 and Q6, to adjust the balance of the Q5/Q6 current mirror. The potentiometer is adjusted such that the output is null (midrange) when the inputs are shorted together.

Non-Linear Characteristics

Input Breakdown Voltage

The transistors Q3, Q4 help to increase the reverse V_{BE} rating: the base-emitter junctions of the NPN transistors Q1 and Q2 break down at around 7V, but the PNP transistors Q3 and Q4 have V_{BE} breakdown voltages around 50 V.

Output-Stage Voltage Swing and Current Limiting

Variations in the quiescent current with temperature, or between parts with the same type number, are common, so crossover distortion and quiescent current may be subject to significant variation.

The output range of the amplifier is about one volt less than the supply voltage, owing in part to V_{BE} of the output transistors Q14 and Q20.

The 25 Ω resistor at the Q14 emitter, along with Q17, acts to limit Q14 current to about 25 mA; otherwise, Q17 conducts no current.

Current limiting for Q20 is performed in the voltage gain stage: Q22 senses the voltage across Q19›s emitter resistor (50Ω); as it turns on, it diminishes the drive current to Q15 base.

Later versions of this amplifier schematic may show a somewhat different method of output current limiting.

Applicability Considerations

The "741" has come to often mean a generic op-amp IC (such as μA741, LM301, 558, LM324, TBA221 — or a more modern replacement such as the TL071). The description of the 741 output stage is qualitatively similar for many other designs (that may have quite different input stages), except:

- Some devices (μA748, LM301, LM308) are not internally compensated (require an external capacitor from output to some point within the operational amplifier, if used in low closed-loop gain applications).
- Some modern devices have rail-to-rail output capability, meaning that the output can range from within a few millivolts of the positive supply voltage to within a few millivolts of the negative supply voltage.

REFERENCES

1. Yuhua Cheng; Chenming Hu (1999). "§2.1 MOSFET classification and operation". MOSFET modeling & BSIM3 user's guide. Springer. p. 13. ISBN 0-7923-8575-6.
2. U. A. Bakshi; A. P. Godse (2007). "§8.2 The depletion mode MOSFET". Electronic Circuits. Technical Publications. pp. 8–2. ISBN 978-81-8431-284-3.
3. Lilienfeld Julius Edgar U.S. Patent 1,745,175 "Method and apparatus for controlling electric currents". Priority date October 22, 1925
4. "1960 – Metal Oxide Semiconductor (MOS) Transistor Demonstrated: John Atalla and Dawon Kahng fabricate working transistors and demonstrate the first successful MOS field-effect amplifier (U.S. Patent 3,102,230 filed in 1960, issued in 1963)". Computer History Museum.
5. "The Silicon Engine | 1948 – Conception of the Junction Transistor". Computer History Museum. 2007.
6. "Intel 45nm Hi-k Silicon Technology". Archived from the original on October 6, 2009.
7. "Using a MOSFET as a Switch". 090507 brunningsoftware.co.uk
8. H Shichman & DA Hodges (1968). "Modeling and simulation of insulated-gate field-effect transistor switching circuits". IEEE Journal of Solid-state circuits. SC–3 (3): 285–289. doi:10.1109/JSSC.1968.1049902. Archived from the original on June 10, 2013. and often is called the Shichman-Hodges model.
9. For example, see Yuhua Cheng; Chenming Hu (1999). MOSFET modeling & BSIM3 user's guide. Springer. ISBN 0-7923-8575-6.. The most recent version of the BSIM model is described in Sriramkumar V.; Navid Paydavosi; Darsen Lu; Chung-Hsun Lin; Mohan Dunga; Shijing Yao; Tanvir Morshed; Ali Niknejad & Chenming Hu (2012). "BSIM-CMG 106.1.0beta Multi-Gate MOSFET Compact Model" (PDF). Department of EE and CS, UC Berkeley. Retrieved 2012-04-01.
10. =P R Gray; P J Hurst; S H Lewis & R G Meyer (2001). Analysis and Design of Analog Integrated Circuits (Fourth ed.). New York: Wiley. pp. 66–67. ISBN 0-471-32168-0.
11. P. R. van der Meer, A. van Staveren, A. H. M. van Roermund (2004). Low-Power Deep Sub-Micron CMOS Logic: Subthreshold Current Reduction. Dordrecht: Springer. p. 78. ISBN 1-4020-2848-2.
12. Degnan, Brian. "Wikipedia fails subvt".
13. Mead, Carver (1989). Analog VLSI and Neural Systems. Reading, MA: Addison-Wesley. p. 370. ISBN 9780201059922.
14. =Leslie S. Smith; Alister Hamilton (1998). Neuromorphic Systems: Engineering Silicon from Neurobiology. World Scientific. pp. 52–56. ISBN 981-02-3377-9.
15. Satish Kumar (2004). Neural Networks: A Classroom Approach. Tata McGraw-Hill. p. 688. ISBN 0-07-048292-6.

16. =Manfred Glesner; Peter Zipf; Michel Renovell (2002). Field-programmable Logic and Applications: 12th International Conference. Dordrecht: Springer. p. 425. ISBN 3-540-44108-5.

17. Eric A. Vittoz (1996). "The Fundamentals of Analog Micropower Design". In Chris Toumazou; Nicholas C. Battersby; Sonia Porta. Circuits and systems tutorials. John Wiley and Sons. pp. 365–372. ISBN 978-0-7803-1170-1.

18. =Sandeep K. Shukla; R. Iris Bahar (2004). Nano, Quantum and Molecular Computing. Springer. p. 10 and Fig. 1.4, p. 11. ISBN 1-4020-8067-0.

19. =Ashish Srivastava; Dennis Sylvester; David Blaauw (2005). Statistical Analysis and Optimization For VLSI: Timing and Power. Springer. p. 135. ISBN 0-387-25738-1.

20. =C Galup-Montoro & Schneider MC (2007). MOSFET modeling for circuit analysis and design. London/Singapore: World Scientific. p. 83. ISBN 981-256-810-7.

21. Norbert R Malik (1995). Electronic circuits: analysis, simulation, and design. Englewood Cliffs, NJ: Prentice Hall. pp. 315–316. ISBN 0-02-374910-5.

22. PR Gray; PJ Hurst; SH Lewis; RG Meyer. §1.5.2 p. 45. ISBN 0-471-32168-0.

23. =A. S. Sedra & K.C. Smith (2004). Microelectronic circuits (Fifth ed.). New York: Oxford. p. 552. ISBN 0-19-514251-9.

24. =A. S. Sedra & K.C. Smith. p. 250, Eq. 4.14. ISBN 0-19-514251-9.

25. Body effect. Equars.com. Retrieved 2012-06-02.

26. Electronic Circuit Symbols. circuitstoday.com. 9 November 2011

27. IEEE Std 315-1975 — Graphic Symbols for Electrical and Electronics Diagrams (Including Reference Designation Letters)

28. Microelectronic Circuit Design, Jaeger, Blalock — Figure 4.15 IEEE Standard MOS transistor circuit symbols.

29. memory components data book (PDF). memory components data book. Intel. pp. 2–1. Retrieved 30 August 2015.

30. Computer History Museum – The Silicon Engine | 1955 – Photolithography Techniques Are Used to Make Silicon Devices. Computerhistory.org. Retrieved 2012-06-02.

31. 1964 - First Commercial MOS IC Introduced.

32. Robert H. Cushman. 2-1/2-generation μP's-$10 parts that perform like low-end minis. EDN September 20, 1975.

33. Computer History Museum – The Silicon Engine | 1963 – Complementary MOS Circuit Configuration is Invented. Computerhistory.org. Retrieved 2012-06-02.

34. Computer History Museum – Exhibits – Microprocessors. Computerhistory.org. Retrieved 2012-06-02.

35. =Jean-Pierre Colinge; Cynthia A. Colinge (2002). Physics of Semiconductor Devices. Dordrecht: Springer. p. 233, Figure 7.46. ISBN 1-4020-7018-7.

36. Eicke R. Weber, Jarek Dabrowski (Editors) (2004). Predictive Simulation of Semiconductor Processing: Status and Challenges. Dordrecht: Springer. p. 5, Figure 1.2. ISBN 3-540-20481-4.

37. International Technology Roadmap for Semiconductors.

38. 1965 – Moore›s Law Predicts the Future of Integrated Circuits. Computer History Museum.

39. =Kaushik Roy; Kiat Seng Yeo (2004). Low Voltage, Low Power VLSI Subsystems. McGraw-Hill Professional. Fig. 2.1, p. 44. ISBN 0-07-143786-X.

40. =Kaushik Roy; Kiat Seng Yeo (2004). Low Voltage, Low Power VLSI Subsystems. McGraw-Hill Professional. Fig. 1.1, p. 4. ISBN 0-07-143786-X.

41. Dragica Vasileska; Stephen Goodnick (2006). Computational Electronics. Morgan & Claypool. p. 103. ISBN 1-59829-056-8.

42. Frontier Semiconductor Paper (PDF). Archived from the original (PDF) on February 27, 2012. Retrieved 2012-06-02.

43. Wai-Kai Chen (2006). The VLSI Handbook. CRC Press. Fig. 2.28, p. 2–22. ISBN 0-8493-4199-X.

44. R. Lindsay; Pawlak; Kittl; Henson; Torregiani; Giangrandi; Surdeanu; Vandervorst; Mayur; Ross; McCoy; Gelpey; Elliott; Pages; Satta; Lauwers; Stolk; Maex (2011). A Comparison of Spike, Flash, SPER and Laser Annealing for 45nm CMOS. MRS Proceedings. 765. doi:10.1557/PROC-765-D7.4.

45. VLSI wiring capacitance[dead link]. IBM Journal of Research and Development.

46. D Soudris, P Pirsch, E Barke (Editors) (2000). Integrated Circuit Design: Power and Timing Modeling, Optimization, and Simulation (10th Int. Workshop). Springer. p. 38. ISBN 3-540-41068-6.

47. Michael Orshansky; Sani Nassif; Duane Boning (2007). Design for Manufacturability And Statistical Design: A Constructive Approach. New York 309284: Springer.

48. PM Zeitzoff; JA Hutchby; HR Huff (2002). Figure 12: Simplified cross section of FinFET double-gate MOSFET.. In Yoon-Soo Park; Michael Shur; William Tang. Frontiers in electronics: future chips : proceedings of the 2002 Workshop on Frontiers in Electronics (WOFE-02), St Croix, Virgin Islands, USA, 6–11 January 2002. World Scientific. p. 82. ISBN 981-238-222-4.

49. J-H Lee; J-W Lee; H-A-R Jung; B-K Choi (2009). Comparison of SOI FinFETs and bulk FinFETs: Figure 2. Silicon-on-Insulator Technology and Devices. The Electrochemical Society. p. 102. ISBN 1-56677-712-7.

50. Techweb. Techweb (2010-01-29). Retrieved 2010-11-27.

51. Power Semiconductor Devices, B. Jayant Baliga, PWS publishing Company, Boston. ISBN 0-534-94098-6

52. Power MOSFET Basics: Understanding MOSFET Characteristics Associated With The Figure of Merit. element14. Retrieved 2010-11-27. Archived April 5, 2015, at the Wayback Machine.

53. Power MOSFET Basics: Understanding Gate Charge and Using It To Assess Switching Performance. element14. Retrieved 2010-11-27. Archived June 30, 2014, at the Wayback Machine.

Feedback Oscillator

The most common form of linear oscillator is an electronic amplifier such as a transistor or op amp connected in a feedback loop with its output fed back into its input through a frequency selective electronic filter to provide positive feedback. When the power supply to the amplifier is first switched on, electronic noise in the circuit provides a non-zero signal to get oscillations started. The noise travels around the loop and is amplified and filtered until very quickly it converges on a sine wave at a single frequency.

Feedback oscillator circuits can be classified according to the type of frequency selective filter they use in the feedback loop:

- In an *RC oscillator* circuit, the filter is a network of resistors and capacitors. RC oscillators are mostly used to generate lower frequencies, for example in the audio range. Common types of RC oscillator circuits are the phase shift oscillator and the Wien bridge oscillator.
- In an *LC oscillator* circuit, the filter is a tuned circuit (often called a *tank circuit*; the tuned circuit is a resonator) consisting of an inductor (L) and capacitor (C) connected together. Charge flows back and forth between the capacitor's plates through the inductor, so the tuned circuit can store electrical energy oscillating at its resonant frequency. There are small losses in the tank circuit, but the amplifier compensates for those losses and supplies the power for the output signal. LC oscillators are often used at radio frequencies, when a tunable frequency source is necessary, such as in signal generators, tunable radio transmitters and the local oscillators in radio receivers. Typical LC oscillator circuits are the Hartley, Colpitts and Clapp circuits.

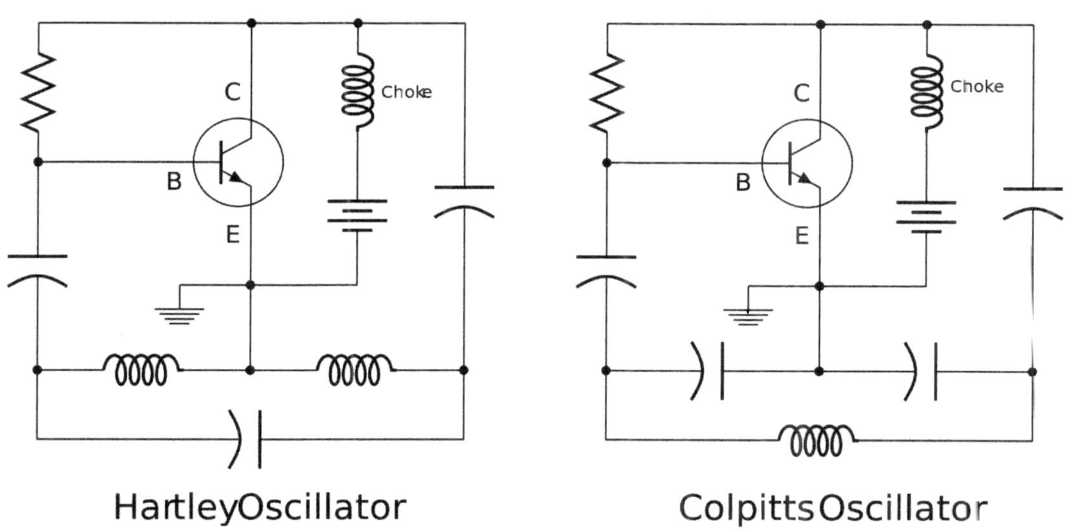

Two common LC oscillator circuits, the Hartley and Colpitts oscillators

- In a *crystal oscillator* circuit the filter is a piezoelectric crystal (commonly a quartz crystal). The crystal mechanically vibrates as a resonator, and its frequency of vibration determines the oscillation frequency. Crystals have very high Q-factor and also better temperature stability than tuned circuits, so crystal oscillators have much better frequency stability than LC or RC oscillators. Crystal oscillators are the most common type of linear oscillator, used to stabilize the frequency of most radio transmitters, and to generate the clock signal in computers and quartz clocks. Crystal oscillators often use the same circuits as LC oscillators, with the crystal replacing the tuned circuit; the Pierce oscillator circuit is also commonly used. Quartz crystals are generally limited to frequencies of 30 MHz or below. Other types of resonator, dielectric resonators and surface acoustic wave (SAW) devices, are used to control higher frequency oscillators, up into the microwave range. For example, SAW oscillators are used to generate the radio signal in cell phones.

Negative Resistance Oscillator

In addition to the feedback oscillators described above, which use two-port amplifying active elements such as transistors and op amps, linear oscillators can also be built using one-port (two terminal) devices with negative resistance, such as magnetron tubes, tunnel diodes, lambda diodes and Gunn diodes. Negative resistance oscillators are usually used at high frequencies in the microwave range and above, since at these frequencies feedback oscillators perform poorly due to excessive phase shift in the feedback path.

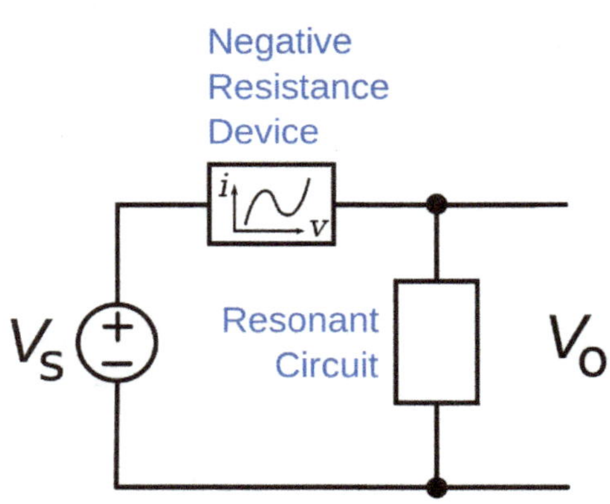

(left) Typical block diagram of a negative resistance oscillator. In some types the negative resistance device is connected in parallel with the resonant circuit. *(right)* A negative resistance microwave oscillator consisting of a Gunn diode in a cavity resonator. The negative resistance of the diode excites microwave oscillations in the cavity, which radiate out the aperture into a waveguide.

In negative resistance oscillators, a resonant circuit, such as an LC circuit, crystal, or cavity resonator, is connected across a device with negative differential resistance, and a DC bias voltage is applied to supply energy. A resonant circuit by itself is "almost" an oscillator; it can store energy in the form of electronic oscillations if excited, but because it has electrical resistance and other losses the oscillations are damped and decay to zero. The negative resistance of the active device cancels the (positive) internal loss resistance in the resonator, in effect creating a resonator with no damping, which generates spontaneous continuous oscillations at its resonant frequency.

The negative resistance oscillator model is not limited to one-port devices like diodes; feedback oscillator circuits with two-port amplifying devices such as transistors and tubes also have negative resistance. At high frequencies, transistors and FETs do not need a feedback loop, but with certain loads applied to one port can become unstable at the other port and show negative resistance due to internal feedback, causing them to oscillate. So high frequency oscillators in general are designed using negative resistance techniques.

Some of the many harmonic oscillator circuits are listed below:

Active devices used in oscillators and approximate maximum frequencies	
Device	**Frequency**
Triode vacuum tube	~1 GHz
Bipolar transistor (BJT)	~20 GHz
Heterojunction Bipolar Transistor (HBT)	~50 GHz
Metal Semiconductor Field Effect Transistor (MESFET)	~100 GHz
Gunn diode, fundamental mode	~100 GHz
Magnetron tube	~100 GHz
High Electron Mobility Transistor (HEMT)	~200 GHz
Klystron tube	~200 GHz
Gunn diode, harmonic mode	~200 GHz
IMPATT diode	~300 GHz
Gyrotron tube	~300 GHz

Armstrong oscillator

- Clapp oscillator
- Colpitts oscillator
- Cross-coupled oscillator
- Dynatron oscillator
- Hartley oscillator
- Meissner oscillator
- Opto-electronic oscillator
- Pierce oscillator
- Phase-shift oscillator
- Robinson oscillator
- Tri-tet oscillator

- Vackář oscillator
- Wien bridge oscillator

RELAXATION OSCILLATOR

A nonlinear or relaxation oscillator produces a non-sinusoidal output, such as a square, sawtooth or triangle wave. It consists of an energy-storing element (a capacitor or, more rarely, an inductor) and a nonlinear switching device (a latch, Schmitt trigger, or negative resistance element) connected in a feedback loop. The switching device periodically charges and discharges the energy stored in the storage element thus causing abrupt changes in the output waveform.

Square-wave relaxation oscillators are used to provide the clock signal for sequential logic circuits such as timers and counters, although crystal oscillators are often preferred for their greater stability. Triangle wave or sawtooth oscillators are used in the timebase circuits that generate the horizontal deflection signals for cathode ray tubes in analogue oscilloscopes and television sets. They are also used in voltage controlled oscillators (VCOs), inverters and switching power supplies, dual slope analog to digital converters (ADCs), and in function generators to generate square and triangle waves for testing equipment. In general, relaxation oscillators are used at lower frequencies and have poorer frequency stability than linear oscillators.

Ring oscillators are built of a ring of active delay stages. Generally the ring has an odd number of inverting stages, so that there is no single stable state for the internal ring voltages. Instead, a single transition propagates endlessly around the ring.

Some of the more common relaxation oscillator circuits are listed below:

- Multivibrator
- Pearson-Anson oscillator
- Ring oscillator
- Delay line oscillator
- Royer oscillator

VOLTAGE-CONTROLLED OSCILLATOR (VCO)

An oscillator can be designed so that the oscillation frequency can be varied over some range by an input voltage or current. These voltage controlled oscillators are widely used in phase-locked loops, in which the oscillator's frequency can be locked to the frequency of another oscillator. These are ubiquitous in modern communications circuits, used in filters, modulators, demodulators, and forming the basis of frequency synthesizer circuits which are used to tune radios and televisions.

Radio frequency VCOs are usually made by adding a varactor diode to the tuned circuit or resonator in an oscillator circuit. Changing the DC voltage across the varactor changes its capacitance, which changes the resonant frequency of the tuned circuit. Voltage controlled relaxation oscillators can be constructed by charging and discharging the energy storage capacitor with a voltage controlled current source. Increasing the input voltage increases the rate of charging the capacitor, decreasing the time between switching events.

HISTORY

One of the first electronic oscillators was an oscillating arc built by Elihu Thomson in 1892. Thomson's oscillator placed an LC tuned circuit in parallel with the arc, used metal electrodes, and included a magnetic blowout. Independently in the same year, George Francis Fitzgerald realized that if the damping resistance in a resonant circuit could be made zero or negative, it would produce oscillations, and tried unsuccessfully to build a negative resistance oscillator with a dynamo, what would now be called a parametric oscillator. The arc oscillator was rediscovered and popularized by William Duddell in 1900. Electric arcs were used to provide illumination in the 19th century, but the arc current was unstable and they often produced hissing, humming or howling sounds. Duddell, a student at London Technical College, investigated this effect. He attached an LC circuit to the electrodes of an arc lamp, and the negative resistance of the arc excited audio frequency oscillations in the tuned circuit at its resonant frequency. Some of the energy was radiated as sound waves by the arc, producing a musical tone. To demonstrate his oscillator before the London Institute of Electrical Engineers, Duddell wired a series of tuned circuits to the arc and played a tune, "*God Save The Queen*". Duddell wasn't able to generate frequencies above the audio range with his "singing arc", but in 1902 Danish physicists Valdemar Poulsen and P. O. Pederson were able to increase the frequency produced into the radio range, inventing the Poulsen arcradio transmitter, the first continuous wave radio transmitter, which was used through the 1920s.

A 120 MHz oscillator from 1938 using a parallel rod transmission line resonator (Lecher line). Transmission lines are widely used for UHF oscillators.

The vacuum tube feedback oscillator was invented around 1912, when it was discovered that feedback ("regeneration") in the recently invented audion vacuum tube could produce oscillations. At least six researchers independently made this discovery and can be said to have some role in the invention. In the summer of 1912, Edwin Armstrong observed oscillations in audion radio receiver circuits and went on to use positive feedback in his invention of the regenerative receiver. German Alexander Meissner independently discovered positive feedback and invented oscillators in March 1913. Irving Langmuir at General Electric observed feedback in 1913. Fritz Lowenstein may have preceded the others with a crude oscillator in late 1911. In Britain, H. J. Round patented amplifying and oscillating circuits in 1913. In August 1912, Lee De Forest, the inventor of the audion, had also observed oscillations in his amplifiers, but he didn't understand its significance and tried to eliminate it until he read Armstrong's patents in 1914, which he promptly challenged. Armstrong and De Forest fought a protracted legal battle over the rights to the "regenerative" oscillator circuit which has been called "the most complicated patent litigation in the history of radio". De Forest ultimately won before the Supreme Court in 1934 on technical grounds, but most sources regard Armstrong's claim as the stronger one.

The first and most widely used relaxation oscillator circuit, the astable multivibrator, was invented in 1917 by French engineers Henri Abraham and Eugene Bloch. They called their cross-coupled, dual vacuum tube circuit a *multivibrateur* because the square-wave signal it produced was rich in harmonics, compared to the sinusoidal signal of other vacuum tube oscillators.

Vacuum tube feedback oscillators became the basis of radio transmission by 1920. However the triode vacuum tube oscillator performed poorly above 300 MHz because of interelectrode capacitance. To reach higher frequencies, new "transit time" (velocity modulation) vacuum tubes were developed, in which electrons traveled in "bunches" through the tube. The first of these was the Barkhausen-Kurz oscillator (1920), the first tube to produce power in the UHF range. The most important and widely used were the klystron (R. and S. Varian, 1937) and the cavity magnetron (J. Randall and H. Boot, 1940).

Mathematical conditions for feedback oscillations, now called the Barkhausen criterion, were derived by Heinrich Georg Barkhausen in 1921. The first analysis of a nonlinear electronic oscillator model, the Van der Pol oscillator, was done by Balthasar van der Pol in 1927. He showed that the stability of the oscillations (limit cycles) in actual oscillators was due to the nonlinearity of the amplifying device. He originated the term "relaxation oscillation" and was first to distinguish between linear and relaxation oscillators. Further advances in mathematical analysis of oscillation were made by Hendrik Wade Bode and Harry Nyquist in the 1930s. In 1969 K. Kurokawa derived necessary and sufficient conditions for oscillation in negative resistance circuits, which form the basis of modern microwave oscillator design.

REFERENCES

1. Snelgrove, Martin (2011). "Oscillator". McGraw-Hill Encyclopedia of Science and Technology, 10th Ed., Science Access online service. McGraw-Hill. Retrieved March 1, 2012.

2. Chattopadhyay, D. (2006). Electronics (fundamentals And Applications). New Age International. pp. 224–225. ISBN 81-224-1780-9.
3. Garg, Rakesh Kumar; Ashish Dixit; Pavan Yadav (2008). Basic Electronics. Firewall Media. p. 280. ISBN 8131803023.
4. Kung, Fabian Wai Lee (2009). "Lesson 9: Oscillator Design" (PDF). RF/ Microwave Circuit Design. Prof. Kung's website, Multimedia University. Retrieved October 17, 2012. External link in |work= (help), Sec. 3 Negative Resistance Oscillators, p. 9-10, 14
5. Räisänen, Antti V.; Arto Lehto (2003). Radio Engineering for Wireless Communication and Sensor Applications. USA: Artech House. pp. 180–182. ISBN 1580535429.
6. Ellinger, Frank (2008). Radio Frequency Integrated Circuits and Technologies, 2nd Ed. USA: Springer. pp. 391–394. ISBN 3540693246.
7. Maas, Stephen A. (2003). Nonlinear Microwave and RF Circuits, 2nd Ed. Artech House. pp. 542–544. ISBN 1580534848.
8. US 500630, Thomson, Elihu, "Method of and Means for Producing Alternating Currents", published 18 July 1892, issued 4 July 1893
9. G. Fitzgerald, On the Driving of Electromagnetic Vibrations by Electromagnetic and Electrostatic Engines, read at the January 22, 1892 meeting of the Physical Society of London, in Larmor, Joseph, ed. (1902). The Scientific Writings of the late George Francis Fitzgerald. London: Longmans, Green and Co. pp. 277–281.
10. Hong, Sungook (2001). Wireless: From Marconi's Black-Box to the Audion. MIT Press. ISBN 0262082985., pp. 161–165
11. Morse 1925, pp. 80–81
12. GB 190021629, Duddell, William du Bois, "Improvements in and connected with Means for the Conversion of Electrical Energy, Derived from a Source of Direct Current, into Varying or Alternating Currents", published 29 Nov 1900, issued 23 Nov 1901
13. Morse 1925, p. 31
14. GB 190315599, Poulsen, "Improvements relating to the Production of Alternating Electric Currents", issued 14 July 1904
15. US 789449, Poulsen, Valdemar, "Method of Producing Alternating Currents with a High Number of Vibrations", issued 9 May 1905
16. Hempstead, Colin; William E. Worthington (2005). Encyclopedia of 20th-Century Technology. 2. Taylor & Francis. p. 648. ISBN 1579584640.
17. Hong 2001, p. 156
18. Fleming, John Ambrose (1919). The Thermionic Valve and its Developments in Radiotelegraphy and Telephony. London: The Wireless Press. pp. 148–155.
19. Hong, Sungook (2003). "A history of the regeneration circuit: From invention to patent litigation" (PDF). IEEE. Retrieved August 29, 2012., pp. 9–10

20. Armstrong, Edwin H. (September 1915). "Some recent developments in the Audion receiver" (PDF). Proc. of the IRE. New York: Institute of Radio Engineers. 3 (9): 215–247. doi:10.1109/jrproc.1915.216677. Retrieved August 29, 2012.
21. Hijiya, James A. (1992). Lee De Forest and the Fatherhood of Radio. Lehigh University Press. pp. 89–90. ISBN 0934223238.
22. Nahin, Paul J. (2001). The Science of Radio: With Matlab and Electronics Workbench Demonstration, 2nd Ed. Springer. p. 280. ISBN 0387951504.
23. Abraham, H.; E. Bloch (1919). "Measurement of period of high frequency oscillations". Comptes Rendus. French Academy of Sciences. 168: 1105.
24. Glazebrook, Richard (1922). A Dictionary of Applied Physics, Vol. 2: Electricity. London: Macmillan and Co. Ltd. pp. 633–634.
25. Calvert, James B. (2002). "The Eccles-Jordan Circuit and Multivibrators". Dr. J. B. Calvert website, Univ. of Denver. Retrieved May 15, 2013. External link in |publisher= (help)
26. Van der Pol, Balthazar (1927). "On relaxation-oscillations". The London, Edinburgh and Dublin Philosophical Magazine. 2 (7): 978–992. doi:10.1080/14786442608564127.
27. Nyquist, H. (January 1932). "Regeneration Theory" (PDF). Bell System Tech. J. USA: American Tel. & Tel. 11 (1): 126–147. doi:10.1002/j.1538-7305.1932.tb02344.x. Retrieved December 5, 2012. on Alcatel-Lucent website
28. Kurokawa, K. (July 1969). "Some Basic Characteristics of Broadband Negative Resistance Oscillator Circuits" (PDF). Bell System Tech. J. USA: American Tel. & Tel. 48 (6): 1937–1955. doi:10.1002/j.1538-7305.1969.tb01158.x. Retrieved December 8, 2012. Eq. 10 is necessary condition for oscillation, eq. 12 is sufficient condition.

CHAPTER 6

Integrated Circuits

Chapter Contents

- **Introduction**
- **Terminology**
- **Invention**
- **Advances**
- **Design**
- **Manufacturing**
- **Intellectual Property**
- **Generations**

An integrated circuit, or IC, is small chip that can function as an amplifier, oscillator, timer, microprocessor, or even computer memory. An IC is a small wafer, usually made of silicon, that can hold anywhere from hundreds to millions of transistors, resistors, and capacitors. These extremely small electronics can perform calculations and store data using either digital or analog technology.

Digital ICs use logic gates, which work only with values of ones and zeros. A low signal sent to a component on a digital IC will result in a value of 0, while a high signal creates a value of 1. Digital ICs are the kind you will usually find in computers, networking equipment, and most consumer electronics.

Analog, or linear ICs work with continuous values. This means a component on a linear IC can take a value of any kind and output another value. The term "linear" is used since the output value is a linear function of the input. For example, a component on a linear IC may multiple an incoming value by a factor of 2.5 and output the result. Linear ICs are typically used in audio and radio frequency amplification.

INTRODUCTION

An integrated circuit or monolithic integrated circuit (also referred to as an IC, a chip, or a microchip) is a set of electronic circuits on one small flat piece (or "chip") of semiconductor material, normally silicon. The integration of large numbers of tiny transistors into a small chip resulted in circuits that are orders of magnitude smaller, cheaper, and faster than those constructed of discrete electronic components. The IC's mass production capability, reliability and building-block approach to circuit design ensured the rapid adoption of standardized ICs in place of designs using discrete transistors. ICs are now used in virtually all electronic equipment and have revolutionized the world of electronics. Computers, mobile phones, and other digital home appliances are now inextricable parts of the structure of modern societies, made possible by the small size and low cost of ICs.

ICs were made possible by experimental discoveries showing that semiconductor devices could perform the functions of vacuum tubes, and by mid-20th-century technology advancements in semiconductor device fabrication. Since their origins in the 1960s, the size, speed, and capacity of chips has increased enormously, driven by technical advances that allow more and more transistors on chips of the same size - a modern chip may have several billion transistors in an area the size of a human fingernail. These advances, roughly following Moore's law, allow a computer chip of 2016 to have millions of times the capacity and thousands of times the speed of the computer chips of the early 1970s.

Erasable programmable read-only memory integrated circuits. These packages have a transparent window that shows the die inside. The window allows the memory to be erased by exposing the chip to ultraviolet light.

Integrated circuit from an EPROM memory microchip showing the memory blocks, the supporting circuitry and the fine silver wires which connect the integrated circuit die to the legs of the packaging.

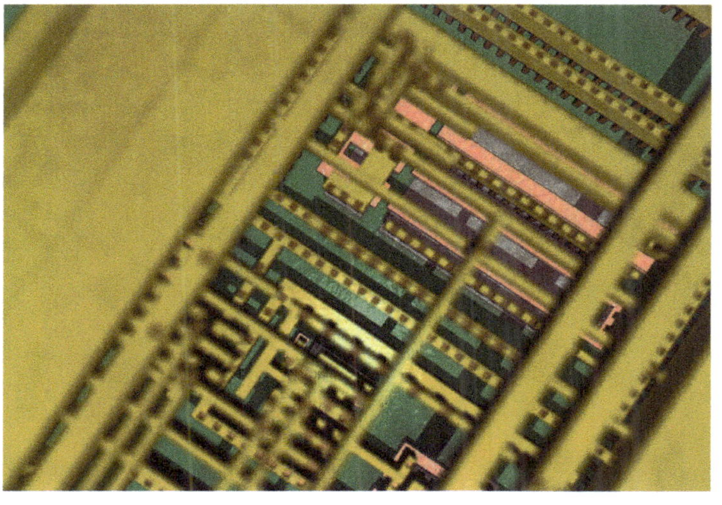

Synthetic detail of an integrated circuit through four layers of planarized copper interconnect, down to the polysilicon (pink), wells (greyish), and substrate (green)

ICs have two main advantages over discrete circuits: cost and performance. Cost is low because the chips, with all their components, are printed as a unit by photolithography rather than being constructed one transistor at a time. Furthermore, packaged ICs use much less material than discrete circuits. Performance is high because the IC's components switch quickly and consume little power (compared to their discrete counterparts) because of their small size and close proximity. The main disadvantage of ICs is the high cost to design them and fabricate the required photomasks. This high initial cost means ICs are only practical when high production volumes are anticipated.

TERMINOLOGY

An *integrated circuit* is defined as:

A circuit in which all or some of the circuit elements are inseparably associated and electrically interconnected so that it is considered to be indivisible for the purposes of construction and commerce.

Circuits meeting this definition can be constructed using many different technologies, including thin-film transistor, thick film technology, or hybrid integrated circuit. However, in general usage *integrated circuit* has come to refer to the single-piece circuit construction originally known as a *monolithic integrated circuit*.

INVENTION

Early developments of the integrated circuit go back to 1949, when German engineer Werner Jacobi (Siemens AG) filed a patent for an integrated-circuit-like semiconductor amplifying device showing five transistors on a common substrate in a 3-stage amplifier arrangement. Jacobi disclosed small and cheap hearing aids as typical industrial applications of his patent. An immediate commercial use of his patent has not been reported.

The idea of the integrated circuit was conceived by Geoffrey W.A. Dummer (1909–2002), a radar scientist working for the Royal Radar Establishment of the British Ministry of Defence. Dummer presented the idea to the public at the Symposium on Progress in Quality Electronic Components in Washington, D.C. on 7 May 1952. He gave many symposia publicly to propagate his ideas, and unsuccessfully attempted to build such a circuit in 1956.

A precursor idea to the IC was to create small ceramic squares (wafers), each containing a single miniaturized component. Components could then be integrated and wired into a bidimensional or tridimensional compact grid. This idea, which seemed very promising in 1957, was proposed to the US Army by Jack Kilby and led to the short-lived Micromodule Program (similar to 1951's Project Tinkertoy). However, as the project was gaining momentum, Kilby came up with a new, revolutionary design: the IC.

Newly employed by Texas Instruments, Kilby recorded his initial ideas concerning the integrated circuit in July 1958, successfully demonstrating the first working integrated example on 12 September 1958. In his patent application of 6 February 1959, Kilby described his new device as "a body of semiconductor material … wherein all the components of the electronic circuit are completely integrated." The first customer for the new invention was the US Air Force.

Kilby won the 2000 Nobel Prize in Physics for his part in the invention of the integrated circuit. His work was named an IEEE Milestone in 2009.

Half a year after Kilby, Robert Noyce at Fairchild Semiconductor developed his own idea of an integrated circuit that solved many practical problems Kilby's had not. Noyce's design was made of silicon, whereas Kilby's chip was made of germanium. Noyce credited Kurt Lehovec of Sprague Electric for the principle of p–n junction isolation, a key concept behind the IC. This isolation allows each transistor to operate independently despite being parts of the same piece of silicon.

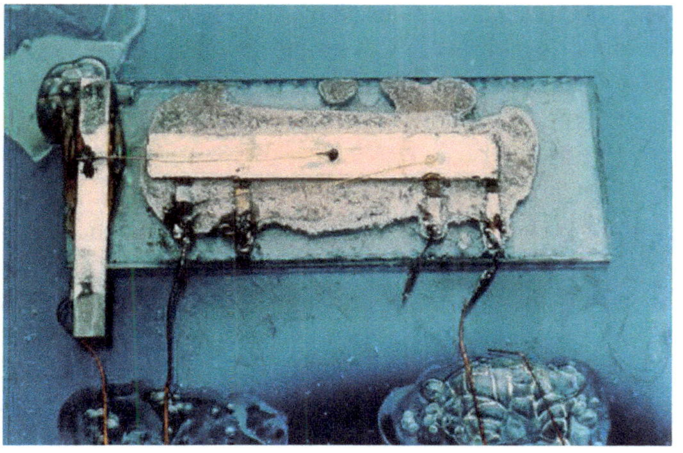

Jack Kilby's original integrated circuit

Fairchild Semiconductor was also home of the first silicon-gate IC technology with self-aligned gates, the basis of all modern CMOS computer chips. The technology was developed by Italian physicist Federico Faggin in 1968, who later joined Intel in order to develop the very first single-chip Central Processing Unit (CPU) (Intel 4004), for which he received the National Medal of Technology and Innovation in 2010.

ADVANCES

Advances in IC technology, primarily smaller features and larger chips, have allowed the number of transistors in an integrated circuit to double every two years, a trend known as Moore's law. This increased capacity has been used to decrease cost and increase functionality. In general, as the feature size shrinks, almost every aspect of an IC's operation improves. The cost per transistor and the switching power consumption per transistor go down, while the memory capacity and speed go up, through the relationships defined by Dennard scaling. Since these speed, capacity, and power consumption gains are apparent to the end user, there is fierce competition among the manufacturers to use finer geometries. Over the years, transistor sizes have decreased from 10s of microns in the early 1970s to around 14 nanometers in 2014 with a corresponding million-fold increase in transistors per unit area. As of 2016, typical chip areas range from a few square millimeters to around 600 mm^2, with up to 25 million transistors per mm^2.

The expected shrinking of feature sizes, and the needed progress in related areas, was forecast for many years by the International Technology Roadmap for Semiconductors (ITRS). The final ITRS was issued in 2016, and it is being replaced by the International Roadmap for Devices and Systems.

Initially, ICs were strictly electronic devices. The success of ICs has led to the integration of other technologies, in the attempt to obtain the same advantages of small size and low cost. These technologies include mechanical devices, optics, and sensors.

- Very small mechanical devices driven by electricity can be integrated onto chips, a technology known as microelectromechanical systems. These devices were developed in the late 1980s and are used in a variety of commercial and military applications. Examples include DLP projectors, inkjet printers, and accelerometers and MEMS gyroscopes used to deploy automobile airbags.
- Since the early 2000s, the integration of optical functionality (optical computing) into silicon chips has been actively pursued in both academic research and in industry resulting in the successful commercialization of silicon based integrated optical transceivers combining optical devices (modulators, detectors, routing) with CMOS based electronics. Integrated optical circuits are also being developed.
- Integrated circuits are also being developed for sensor applications in medical implants or other bioelectronic devices. Special sealing techniques have to be applied in such biogenic environments to avoid corrosion or biodegradation of the exposed semiconductor materials.

As of 2016, the vast majority of all transistors are fabricated in a single layer on one side of a chip of silicon in a flat 2-dimensional planar process. Researchers have produced prototypes of several promising alternatives, such as:

- various approaches to stacking several layers of transistors to make a three-dimensional integrated circuit, such as through-silicon via, "monolithic 3D", stacked wire bonding, etc.
- transistors built from other materials: graphene transistors, molybdenite transistors, carbon nanotube field-effect transistor, gallium nitride transistor, transistor-like nanowire electronic devices, organic field-effect transistor, etc.
- fabricating transistors over the entire surface of a small sphere of silicon.
- modifications to the substrate, typically to make "flexible transistors" for a flexible display or other flexible electronics, possibly leading to a roll-away computer.

DESIGN

The cost of designing and developing a complex integrated circuit is quite high, normally in the multiple tens of millions of dollars. This only makes economic sense if production volume is high, so the non-recurring engineering (NRE) costs are spread across typically millions of production units.

Electronic design automation and Hardware description language

Modern semiconductor chips have billions of components, and are too complex to be designed by hand. Software tools to help the designer are essential. Electronic Design Automation (EDA), also referred to as Electronic Computer-Aided Design (ECAD), is a category of software tools for designing electronic systems, including integrated circuits. The tools work together in a design flow that engineers use to design and analyze entire semiconductor chips.

TYPES

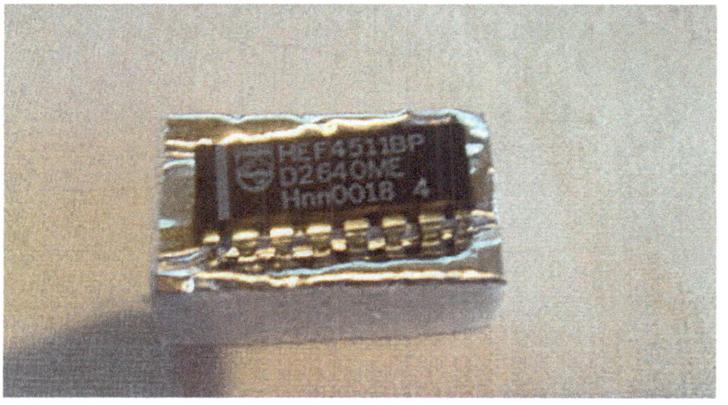

A CMOS 4511 IC in a DIP

Integrated circuits can be classified into analog, digital and mixed signal (both analog and digital on the same chip).

Digital integrated circuits can contain anywhere from one to billions of logic gates, flip-flops, multiplexers, and other circuits in a few square millimeters. The small size of these circuits allows high speed, low power dissipation, and reduced manufacturing cost compared with board-level integration. These digital ICs, typically microprocessors, DSPs, and microcontrollers, work using binary mathematics to process "one" and "zero" signals.

The die from an Intel 8742, an 8-bit microcontroller that includes a CPU running at 12 MHz, 128 bytes of RAM, 2048 bytes of EPROM, and I/O in the same chip

Among the most advanced integrated circuits are the microprocessors or "**cores**", which control everything from computers and cellular phones to digital microwave ovens. Digital memory chips and application-specific integrated circuits (ASICs) are examples of other families of integrated circuits that are important to the modern information society.

In the 1980s, programmable logic devices were developed. These devices contain circuits whose logical function and connectivity can be programmed by the user, rather than being fixed by the integrated circuit manufacturer. This allows a single chip to be programmed to implement different LSI-type functions such as logic gates, adders and registers. Current devices called field-programmable gate arrays (FPGAs) can (as of 2016) implement the equivalent of millions of gates in parallel and operate up to 1 GHz.

Analog ICs, such as sensors, power management circuits, and operational amplifiers, work by processing continuous signals. They perform functions like amplification, active filtering, demodulation, and mixing. Analog ICs ease the burden on circuit designers by having expertly designed analog circuits available instead of designing a difficult analog circuit from scratch.

ICs can also combine analog and digital circuits on a single chip to create functions such as A/D converters and D/A converters. Such mixed-signal circuits offer smaller size and lower cost, but must carefully account for signal interference. Prior to the late 1990s, radios could not be fabricated in the same low-cost CMOS processes as microprocessors. But since 1998, a large number of radio chips have been developed using CMOS processes. Examples include Intel's DECT cordless phone, or 802.11 (Wi-Fi) chips created by Atheros and other companies.

Modern electronic component distributors often further sub-categorize the huge variety of integrated circuits now available:

- Digital ICs are further sub-categorized as logic ICs, memory chips, interface ICs (level shifters, serializer/deserializer, etc.), Power Management ICs, and programmable devices.
- Analog ICs are further sub-categorized as linear ICs and RF ICs.
- mixed-signal integrated circuits are further sub-categorized as data acquisition ICs (including A/D converters, D/A converter, digital potentiometers) and clock/ timing ICs.

MANUFACTURING

Fabrication

Rendering of a small standard cell with three metal layers (dielectric has been removed). The sand-colored structures are metal interconnect, with the vertical pillars being contacts, typically plugs of tungsten. The reddish structures are polysilicon gates, and the solid at the bottom is the crystalline silicon bulk.

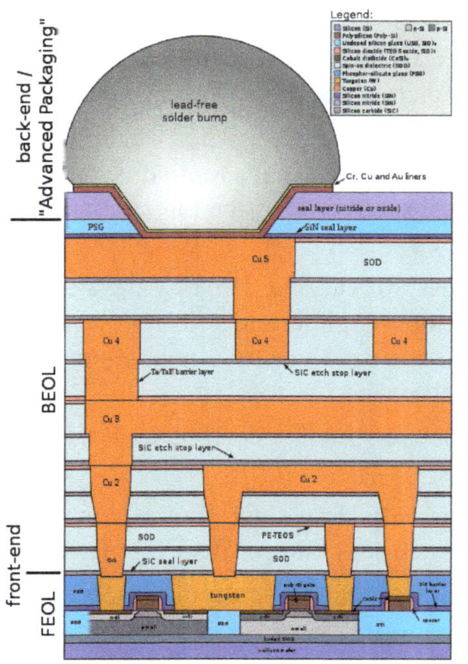

Schematic structure of a CMOS chip, as built in the early 2000s. The graphic shows LDD-MISFET's on an SOI substrate with five metallization layers and solder bump for flip-chip bonding. It also shows the section for FEOL (front-end of line), BEOL (back-end of line) and first parts of back-end process.

The semiconductors of the periodic table of the chemical elements were identified as the most likely materials for a *solid-state vacuum tube*. Starting with copper oxide, proceeding to germanium, then silicon, the materials were systematically studied in the 1940s and 1950s. Today, monocrystalline silicon is the main substrate used for ICs although some III-V compounds of the periodic table such as gallium arsenide are used for specialized applications like LEDs, lasers, solar cells and the highest-speed integrated circuits. It took decades to perfect methods of creating crystals without defects in the crystalline structure of the semiconducting material.

Semiconductor ICs are fabricated in a planar process which includes three key process steps – imaging, deposition and etching. The main process steps are supplemented by doping and cleaning.

Mono-crystal silicon wafers (or for special applications, silicon on sapphire or gallium arsenide wafers) are used as the *substrate*. Photolithography is used to mark different areas of the substrate to be doped or to have polysilicon, insulators or metal (typically aluminium or copper) tracks deposited on them.

- Integrated circuits are composed of many overlapping layers, each defined by photolithography, and normally shown in different colors. Some layers mark where various dopants are diffused into the substrate (called diffusion layers), some define where additional ions are implanted (implant layers), some define the conductors (polysilicon or metal layers), and some define the connections between the conducting layers (via or contact layers). All components are constructed from a specific combination of these layers.
- In a self-aligned CMOS process, a transistor is formed wherever the gate layer (polysilicon or metal) crosses a diffusion layer.
- Capacitive structures, in form very much like the parallel conducting plates of a traditional electrical capacitor, are formed according to the area of the "plates", with insulating material between the plates. Capacitors of a wide range of sizes are common on ICs.
- Meandering stripes of varying lengths are sometimes used to form on-chip resistors, though most logic circuits do not need any resistors. The ratio of the length of the resistive structure to its width, combined with its sheet resistivity, determines the resistance.
- More rarely, inductive structures can be built as tiny on-chip coils, or simulated by gyrators.

Since a CMOS device only draws current on the *transition* between logic states, CMOS devices consume much less current than bipolar devices.

A random-access memory is the most regular type of integrated circuit; the highest density devices are thus memories; but even a microprocessor will have memory on the chip. Although the structures are intricate – with widths which have been shrinking for decades – the layers remain much thinner than the device widths. The layers of material are fabricated much like a photographic process, although light waves in the visible spectrum cannot be used to "expose" a layer of material, as they would be too large for the features. Thus photons of higher frequencies (typically ultraviolet) are used to create the patterns for each layer. Because each feature is so small, electron microscopes are essential tools for a process engineer who might be debugging a fabrication process.

Each device is tested before packaging using automated test equipment (ATE), in a process known as wafer testing, or wafer probing. The wafer is then cut into rectangular blocks, each of which is called a *die*. Each good die (plural *dice*, *dies*, or *die*) is then connected into a package using aluminium (or gold) bond wires which are thermosonically bonded to *pads*, usually found around the edge of the die. . Thermosonic bonding was first introduced by A. Coucoulas which provided a reliable means of forming these vital electrical connections to the outside world. After packaging, the devices go through final testing on the same or similar ATE used during wafer probing. Industrial CT scanning can also be used. Test cost can account for over 25% of the cost of fabrication on lower-cost products, but can be negligible on low-yielding, larger, or higher-cost devices.

As of 2016, a fabrication facility (commonly known as a *semiconductor fab*) can cost over US$8 billion to construct. The cost of a fabrication facility rises over time (Rock's law) because much of the operation is automated. Today, the most advanced processes employ the following techniques:

- The wafers are up to 300 mm in diameter (wider than a common dinner plate).
- As of 2016, a state of the art foundry can produce 14 nm transistors, as implemented by Intel, TSMC, Samsung, and Global Foundries. The next step, to 10 nm devices, is expected in 2017.
- Copper interconnects where copper wiring replaces aluminium for interconnects.
- Low-K dielectric insulators.
- Silicon on insulator (SOI).
- Strained silicon in a process used by IBM known as strained silicon directly on insulator (SSDOI).
- Multigate devices such as tri-gate transistors being manufactured by Intel from 2011 in their 22 nm process.

Packaging

The earliest integrated circuits were packaged in ceramic flat packs, which continued to be used by the military for their reliability and small size for many years. Commercial circuit packaging quickly moved to the dual in-line package (DIP), first in ceramic and later in plastic. In the 1980s pin counts of VLSI circuits exceeded the practical limit for DIP packaging, leading to pin grid array (PGA) and leadless chip carrier (LCC) packages. Surface mount packaging appeared in the early 1980s and became popular in the late 1980s, using finer lead pitch with leads formed as either gull-wing or J-lead, as exemplified by the small-outline integrated circuit (SOIC) package – a carrier which occupies an area about 30–50% less than an equivalent DIP and is typically 70% thinner. This package has "gull wing" leads protruding from the two long sides and a lead spacing of 0.050 inches.

A Soviet MSI nMOS chip made in 1977, part of a four-chip calculator set designed in 1970

In the late 1990s, plastic quad flat pack (PQFP) and thin small-outline package (TSOP) packages became the most common for high pin count devices, though PGA packages are still often used for high-end microprocessors. Intel and AMD are currently transitioning from PGA packages on high-end microprocessors to land grid array (LGA) packages.

Ball grid array (BGA) packages have existed since the 1970s. Flip-chip Ball Grid Array packages, which allow for much higher pin count than other package types, were developed in the 1990s. In an FCBGA package the die is mounted upside-down (flipped) and connects to the package balls via a package substrate that is similar to a printed-circuit board rather than by wires. FCBGA packages allow an array of input-output signals (called Area-I/O) to be distributed over the entire die rather than being confined to the die periphery.

Traces going out of the die, through the package, and into the printed circuit board have very different electrical properties, compared to on-chip signals. They require special design techniques and need much more electric power than signals confined to the chip itself.

When multiple dies are put in one package, the result is a System in Package, or SiP. A Multi-Chip Module, or MCM, is created by combining multiple dies on a small substrate often made of ceramic. The distinction between a big MCM and a small printed circuit board is sometimes fuzzy.

Chip Labeling and Manufacture Date

Most integrated circuits are large enough to include identifying information. Four common sections are the manufacturer's name or logo, the part number, a part production batch number and serial number, and a four-digit date-code to identify when the chip was manufactured. Extremely small surface mount technology parts often bear only a number used in a manufacturer's lookup table to find the chip characteristics.

The manufacturing date is commonly represented as a two-digit year followed by a two-digit week code, such that a part bearing the code 8341 was manufactured in week 41 of 1983, or approximately in October 1983.

INTELLECTUAL PROPERTY

The possibility of copying by photographing each layer of an integrated circuit and preparing photomasks for its production on the basis of the photographs obtained is a reason for the introduction of legislation for the protection of layout-designs. The Semiconductor Chip Protection Act of 1984 established intellectual property protection for photomasks used to produce integrated circuits.

A diplomatic conference was held at Washington, D.C., in 1989, which adopted a Treaty on Intellectual Property in Respect of Integrated Circuits (IPIC Treaty).

The Treaty on Intellectual Property in respect of Integrated Circuits, also called Washington Treaty or IPIC Treaty (signed at Washington on 26 May 1989) is currently not in force, but was partially integrated into the TRIPS agreement.

National laws protecting IC layout designs have been adopted in a number of countries, including Japan, the EC, the UK, Australia, and Korea.

Other Developments

Future developments seem to follow the multi-core multi-microprocessor paradigm, already used by Intel and AMD multi-core processors. Rapport Inc. and IBM started shipping the KC256 in 2006, a 256-core microprocessor. Intel, as recently as February–August 2011, unveiled a prototype, "not for commercial sale" chip that bears 80 cores. Each core is capable of handling its own task independently of the others. This is in response to the heat-versus-speed limit that is about to be reached using existing transistor technology (see: thermal design power). This design provides a new challenge to chip programming. Parallel programming languages such as the open-source X10 programming language are designed to assist with this task.

GENERATIONS

In the early days of simple integrated circuits, the technology's large scale limited each chip to only a few transistors, and the low degree of integration meant the design process was relatively simple. Manufacturing yields were also quite low by today's standards. As the technology progressed, millions, then billions of transistors could be placed on one chip, and good designs required thorough planning, giving rise to the field of Electronic Design Automation, or EDA.

Name	Signification	Year	Transistors number	Logic gates number
SSI	*small-scale integration*	1964	1 to 10	1 to 12
MSI	*medium-scale integration*	1968	10 to 500	13 to 99

LSI	*large-scale integration*	1971	500 to 20,000	100 to 9,999
VLSI	*very large-scale integration*	1980	20,000 to 1,000,000	10,000 to 99,999
ULSI	*ultra-large-scale integration*	1984	1,000,000 and more	100,000 and more

SSI, MSI and LSI

The first integrated circuits contained only a few transistors. Early digital circuits containing tens of transistors provided a few logic gates, and early linear ICs such as the Plessey SL201 or the Philips TAA320 had as few as two transistors. The number of transistors in an integrated circuit has increased dramatically since then. The term "large scale integration" (LSI) was first used by IBM scientist Rolf Landauer when describing the theoretical concept; that term gave rise to the terms "small-scale integration" (SSI), "medium-scale integration" (MSI), "very-large-scale integration" (VLSI), and "ultra-large-scale integration" (ULSI). The early integrated circuits were SSI.

SSI circuits were crucial to early aerospace projects, and aerospace projects helped inspire development of the technology. Both the Minuteman missile and Apollo program needed lightweight digital computers for their inertial guidance systems. Although the Apollo guidance computer led and motivated integrated-circuit technology, it was the Minuteman missile that forced it into mass-production. The Minuteman missile program and various other Navy programs accounted for the total $4 million integrated circuit market in 1962, and by 1968, U.S. Government space and defense spending still accounted for 37% of the $312 million total production. The demand by the U.S. Government supported the nascent integrated circuit market until costs fell enough to allow firms to penetrate the industrial, and eventually, the consumer markets. The average price per integrated circuit dropped from $50.00 in 1962 to $2.33 in 1968. Integrated circuits began to appear in consumer products by the turn of the decade, a typical application being FM inter-carrier sound processing in television receivers.

The first MOS chips were small-scale integration chips for NASA satellites.

The next step in the development of integrated circuits, taken in the late 1960s, introduced devices which contained hundreds of transistors on each chip, called "medium-scale integration" (MSI).

In 1964, Frank Wanlass demonstrated a single-chip 16-bit shift register he designed, with an incredible (at the time) 120 transistors on a single chip.

MSI devices were attractive economically because while they cost little more to produce than SSI devices, they allowed more complex systems to be produced using smaller circuit boards, less assembly work (because of fewer separate components), and a number of other advantages.

Further development, driven by the same economic factors, led to "large-scale integration" (LSI) in the mid-1970s, with tens of thousands of transistors per chip.

The masks used to process and manufacture SSI, MSI and early LSI and VLSI devices (such as the microprocessors of the early 1970s) were mostly created by hand, often using Rubylith-tape or similar. For large or complex ICs (such as memories or processors), this was often done by specially hired layout people under supervision of a team of engineers, who would also, along with the circuit designers, inspect and verify the correctness and completeness of each mask. However, modern VLSI devices contain so many transistors, layers, interconnections, and other features that it is no longer feasible to check the masks or do the original design by hand. The engineer depends on computer programs and other hardware aids to do most of this work.

Integrated circuits such as 1K-bit RAMs, calculator chips, and the first microprocessors, that began to be manufactured in moderate quantities in the early 1970s, had under 4000 transistors. True LSI circuits, approaching 10,000 transistors, began to be produced around 1974, for computer main memories and second-generation microprocessors.

Some SSI and MSI chips, like discrete transistors, are still mass-produced, both to maintain old equipment and build new devices that require only a few gates. The 7400 series of TTL chips, for example, has become a de facto standard and remains in production.

VLSI

The final step in the development process, starting in the 1980s and continuing through the present, was "very-large-scale integration" (VLSI). The development started with hundreds of thousands of transistors in the early 1980s, and continues beyond ten billion transistors as of 2016.

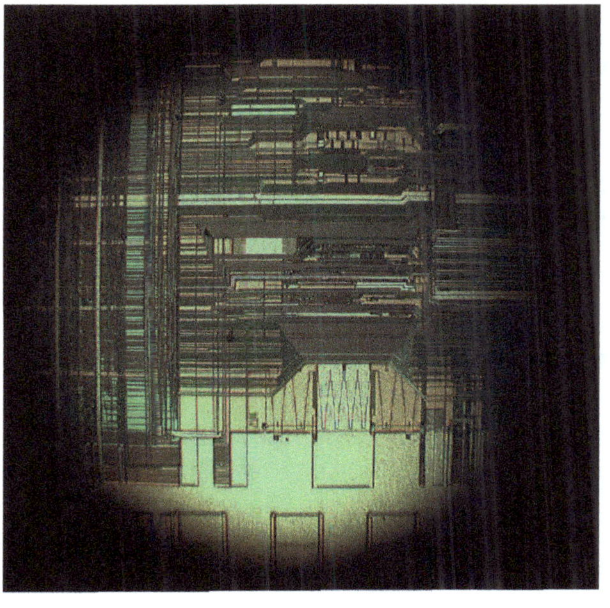

Upper interconnect layers on an Intel 80486DX2 microprocessor die

Multiple developments were required to achieve this increased density. Manufacturers moved to smaller design rules and cleaner fabrication facilities, so that they could make chips with more transistors and maintain adequate yield. The path of process improvements was summarized by the International Technology Roadmap for Semiconductors (ITRS). Design tools improved enough to make it practical to finish these designs in a reasonable time. The more energy-efficient CMOS replaced NMOS and PMOS, avoiding a prohibitive increase in power consumption.

In 1986 the first one-megabit RAM chips were introduced, containing more than one million transistors. Microprocessor chips passed the million-transistor mark in 1989 and the billion-transistor mark in 2005. The trend continues largely unabated, with chips introduced in 2007 containing tens of billions of memory transistors.

ULSI, WSI, SOC and 3D-IC

To reflect further growth of the complexity, the term *ULSI* that stands for "ultra-large-scale integration" was proposed for chips of more than 1 million transistors.

Wafer-scale integration (WSI) is a means of building very large integrated circuits that uses an entire silicon wafer to produce a single "super-chip". Through a combination of large size and reduced packaging, WSI could lead to dramatically reduced costs for some systems, notably massively parallel supercomputers. The name is taken from the term Very-Large-Scale Integration, the current state of the art when WSI was being developed.

A system-on-a-chip (SoC or SOC) is an integrated circuit in which all the components needed for a computer or other system are included on a single chip. The design of such a device can be complex and costly, and building disparate components on a single piece of silicon may compromise the efficiency of some elements. However, these drawbacks are offset by lower manufacturing and assembly costs and by a greatly reduced power budget: because signals among the components are kept on-die, much less power is required.

A three-dimensional integrated circuit (3D-IC) has two or more layers of active electronic components that are integrated both vertically and horizontally into a single circuit. Communication between layers uses on-die signaling, so power consumption is much lower than in equivalent separate circuits. Judicious use of short vertical wires can substantially reduce overall wire length for faster operation.

Silicon Labelling and Graffiti

To allow identification during production most silicon chips will have a serial number in one corner. It is also common to add the manufacturer's logo. Ever since ICs were created, some chip designers have used the silicon surface area for surreptitious, non-functional images or words. These are sometimes referred to as chip art, silicon art, silicon graffiti or silicon doodling.

REFERENCES

1. "Integrated circuit (IC)". JEDEC.
2. Andrew Wylie (2009). "The first monolithic integrated circuits". Retrieved 14 March 2011.
3. Horowitz, Paul; Hill, Winfield (1989). The Art of Electronics (2nd ed.). Cambridge

University Press. p. 61. ISBN 0-521-37095-7.

4. "Integrated circuits help Invention". Integratedcircuithelp.com. Retrieved 2012-08-13.
5. DE 833366 W. Jacobi/SIEMENS AG: "Halbleiterverstärker" priority filing on 14 April 1949, published on 15 May 1952.
6. "The Hapless Tale of Geoffrey Dummer", (n.d.), (HTML), Electronic Product News, accessed 8 July 2008.
7. George Rostky, (n. d.), "Micromodules: the ultimate package", (HTML), EE Times, accessed 8 July 2008.
8. The Chip that Jack Built, (c. 2008), (HTML), Texas Instruments, Retrieved 29 May 2008.
9. Jack S. Kilby, Miniaturized Electronic Circuits, United States Patent Office, US Patent 3,138,743, filed 6 February 1959, issued 23 June 1964.
10. Winston, Brian (1998). Media Technology and Society: A History : From the Telegraph to the Internet. Routledge. p. 221. ISBN 978-0-415-14230-4.
11. "Texas Instruments – 1961 First IC-based computer". Ti.com. Retrieved 2012-08-13.
12. Nobel Web AB, (10 October 2000),(The Nobel Prize in Physics 2000, Retrieved 29 May 2008
13. "Milestones:First Semiconductor Integrated Circuit (IC), 1958". IEEE Global History Network. IEEE. Retrieved 3 August 2011.
14. Kurt Lehovec's patent on the isolation p–n junction: U.S. Patent 3,029,366 granted on 10 April 1962, filed 22 April 1959. Robert Noyce credits Lehovec in his article – "Microelectronics", Scientific American, September 1977, Volume 23, Number 3, pp. 63–9.
15. Davari, Bijan, Robert H. Dennard, and Ghavam G. Shahidi (1995). "CMOS scaling for high performance and low power-the next ten years" (PDF). Proceedings of the IEEE. 83 (4). pp. 595–606.
16. "Ultra-fast, energy-sipping devices powered by Intel". Intel.
17. Hachman, Mark (July 15, 2014). "Intel shipping Broadwell, but next-gen Skylake chip could slip". PC World.
18. "Inside Pascal: NVIDIA's Newest Computing Platform".. 15,300,000,000 transistors in 610 mm2.
19. "International Roadmap for Devices and Systems" (PDF). IEEE. 2016.
20. H. Fujita (1997). A decade of MEMS and its future. Tenth Annual International Workshop on Micro Electro Mechanical Systems.
21. A. Narasimha; et al. (2008). "A 40-Gb/s QSFP optoelectronic transceiver in a 0.13 µm CMOS silicon-on-insulator technology. Proceedings of the Optical Fiber Communication Conference (OFC): OMK7.
22. M. Birkholz; A. Mai; C. Wenger; C. Meliani; R. Scholz (2016). Technology modules from micro- and nano-electronics for the life sciences. WIREs Nanomed.

Nanobiotech. 8: 355–377. doi:10.1002/wnan.1367.

23. A.H.D. Graham; J. Robbins; C.R. Bowen; J. Taylor (2011). Commercialisation of CMOS Integrated Circuit Technology in Multi-Electrode Arrays for Neuroscience and Cell-Based Biosensors. Sensors. 11: 4943–4971. doi:10.3390/s110504943.
24. Zvi Or-Bach. Why SOI is the Future Technology of Semiconductors. 2013.
25. Samsung's Eight-Stack Flash Shows up in Apple's iPhone 4. 2010.
26. Spherical semiconductor radio temperature sensor. Naturelnterface. 2002.
27. Takeda, Nobuo, MEMS applications of Ball Semiconductor Technology (PDF), archived from the original (PDF) on 2015-01-01
28. Mark LaPedus (16 April 2015). FinFET Rollout Slower Than Expected. Semiconductor Engineering.
29. About the EDA Industry. Electronic Design Automation Consortium. Retrieved 29 July 2015.
30. Paul R. Gray; Paul J. Hurst; Stephen H. Lewis; Robert G. Meyer (2009). Analysis and Design of Analog Integrated Circuits. Wiley. ISBN 978-0470245996.
31. Jan M. Rabaey; Anantha Chandrakasan; Borivoje Nikolic (2003). Digital Integrated Circuits (2nd Edition). Pearson. ISBN 978-0130909961.
32. Jacob Baker (2008). CMOS: Mixed-Signal Circuit Design. Wiley. ISBN 978-0470290262.
33. CD4068 data sheet (PDF). Intersil.
34. Stratix 10 Device Overview (PDF). Altera. 12 December 2015. Retrieved 18 Nov 2016.
35. Nathawad, L.; Zargari, M.; Samavati, H.; Mehta, S.; Kheirkhaki, A.; Chen, P.; Gong, K.; Vakili-Amini, B.; Hwang, J.; Chen, M.; Terrovitis, M.; Kaczynski, B.; Limotyrakis, S.; Mack, M.; Gan, H.; Lee, M.; Abdollahi-Alibeik, B.; Baytekin, B.; Onodera, K.; Mendis, S.; Chang, A.; Jen, S.; Su, D.; Wooley, B. 20.2: A Dual-band CMOS MIMO Radio SoC for IEEE 802.11n Wireless LAN (PDF). IEEE Entity Web Hosting. IEEE. Retrieved 22 October 2016.
36. Coucoulas, A., http://commons.wikimedia.org/wiki/File:Hot_Work_Ultrasonic_(Thermosonic)_Bonding_549-556.pdf Hot Work Ultrasonic Bonding – A Method Of Facilitating Metal Flow By Restoration Processes, Proc. 20th IEEE Electronic Components Conf. Washington, D.C., May 1970, pp. 549–556.https://sites.google.com/site/hotworkultrasonicbonding/
37. Max Chafkin; Ian King (June 9, 2016). How Intel Makes a Chip. Bloomburg Businessweek.
38. Mark Lapedus (May 21, 2015). 10 nm Fab Watch. Semiconductor Engineering.
39. 145 series ICs (in Russian). Retrieved 22 April 2012.
40. Federal Statutory Protection for Mask Works (PDF). United States Copyright Office. United States Copyright Office. Retrieved 22 October 2016.
41. On Jan. 1, 1995, the Agreement on Trade-Related Aspects of Intellectual Property Rights (TRIPs) (Annex 1C to the World Trade Organization (WTO) Agreement),

went into force. Part II, section 6 of TRIPs protects semiconductor chip products and was the basis for Presidential Proclamation No. 6780, March 23, 1995, under SCPA § 902(a)(2), extending protection to all present and future WTO members.

42. Japan was the first country to enact its own version of the SCPA, the Japanese Act Concerning the Circuit Layout of a Semiconductor Integrated Circuit of 1985.

43. In 1986 the EC promulgated a directive requiring its members to adopt national legislation for the protection of semiconductor topographies. Council Directive 1987/54/EEC of 16 Dec. 1986 on the Legal Protection of Topographies of Semiconductor Products, art. 1(1)(b), 1987 O.J. (L 24) 36.

44. The UK enacted the Copyright, Designs and Patents Act, 1988, c. 48, § 213, after it initially took the position that its copyright law fully protected chip topographies. See British Leyland Motor Corp. v. Armstrong Patents Co. Criticisms of inadequacy of the UK copyright approach as perceived by the US chip industry are summarized in Further chip rights developments, Micro Law, IEEE Micro, Aug. 1985, pp. 91-92. Australia passed the Circuit Layouts Act of 1989 as a sui generis form of chip protection. Korea passed the Act Concerning the Layout-Design of Semiconductor Integrated Circuits

45. Biever, C. Chip revolution poses problems for programmers, New Scientist (Vol 193, Number 2594)

46. Peter Clarke, Intel enters billion-transistor processor era, EE Times, 14 October 2005

47. http://www.iutbayonne.univ-pau.fr/~dalmau/documents/cours/archi/MICROPancien.pdf

48. Bulletin de la Societe fribourgeoise des sciences naturelles, Volumes 62 à 63 (in French). 1973.

49. Mindell, David A. (2008). Digital Apollo: Human and Machine in Spaceflight. The MIT Press. ISBN 978-0-262-13497-2.

50. Ginzberg, Eli (1976). Economic impact of large public programs: the NASA Experience. Olympus Publishing Company. p. 57. ISBN 0-913420-68-9.

51. Bob Johnstone (1999). We were burning: Japanese entrepreneurs and the forging of the electronic age. Basic Books. pp. 47–48. ISBN 978-0-465-09118-8.

52. Lee Boysel (2007-10-12). Making Your First Million (and other tips for aspiring entrepreneurs). U. Mich. EECS Presentation / ECE Recordings.

53. Intel›s Accidental Revolution. CNET.

54. C. F. O›Donnell. Engineering for systems using large scale integration. p. 870.

55. Peter Clarke, EE Times: Intel enters billion-transistor processor era, 14 November 2005

56. Antone Gonsalves, EE Times, Samsung begins production of 16-Gb flash, 30 April 2007

57. Meindl, J.D. Ultra-large scale integration. ieee.org. IEEE. Retrieved 21 September 2014.

58. Shanefield, Daniel. Wafer scale integration. google.com/patents. Retrieved 21 September 2014.

59. Klaas, Jeff. System-on-a-chip. google.com/patents. Retrieved 21 September 2014.

60. Topol, A.W.; Tulipe, D.C.La; Shi, L; et., al. Three-dimensional integrated circuits. ieee.org. International Business Machines Corporation (IBM). Retrieved 21 September 2014.

CHAPTER 7

Telecommunication Systems

Chapter Contents

- **Introduction**
- **Etymology**
- **History**
- **Key Concepts**
- **Society**
- **Government**
- **Modern Media**
- **Transmission Capacity**

Telecommunication systems transmit data or signals over short or long distances between networks for the usage of communication. Though telecommunications systems did not work as well in the past, there have been vast improvements and is arguably the staple of modern business and society today. Now, more than ever, businesses are making the move towards a network that reach more consumers globally and there are many choices to choose from. These choices include voice over internet protocol (VoIP), and radio, wireless, and radio frequency identification (RFID).

Telecommunication systems only require three elements:

- Information that is taken in by a transmitter and turned into a signal.
- The signal is then carried out through a network medium.
- Signal is then received and translated back into information.

An example of these three telecommunication system elements is how television works. A network station broadcasts shows and commercials and then that signal is carried through cables and into the houses, and it finally gets transmitted and displayed on the television.

Telecommunication systems are usually two-way. An example is point to point communication, which happens through a telephone where there is only one transmitter and receiver.

When more receivers want to be reached than a broadcast communication system is used.

Signals of communication come in two forms: digital or analog. Though analog is still in use more transmitter providers are making the switch towards a digital signal because the information received is much clearer than analog, which is more likely to be interrupted by noise.

Telecommunications systems of today are transferred through many mediums such as the telephone, radio, television, LAN networks and the ever-growing internet. A combination of any form of telecommunication has made businesses more efficient, it has made communication easier, and has globally connected everyone together.

INTRODUCTION

Telecommunication is the transmission of signs, signals, messages, writings, images and sounds or intelligence of any nature by wire, radio, optical or other electromagnetic systems. Telecommunication occurs when the exchange of information between communication participants includes the use of technology. It is transmitted either electrically over physical media, such as cables, or via electromagnetic radiation. Such transmission paths are often divided into communication channels which afford the advantages of multiplexing. The term is often used in its plural form, telecommunications, because it involves many different technologies.

Early means of communicating over a distance included visual signals, such as beacons, smoke signals, semaphore telegraphs, signal flags, and optical heliographs. Other examples of pre-modern long-distance communication included audio messages such as coded drumbeats, lung-blown horns, and loud whistles. 20th and 21st century technologies for long-distance communication usually involve electrical and electromagnetic technologies, such as telegraph, telephone, and teleprinter, networks, radio, microwave transmission, fiber optics, and communications satellites.

A revolution in wireless communication began in the first decade of the 20th century with the pioneering developments in radio communications by Guglielmo Marconi, who won the Nobel Prize in Physics in 1909. Other notable pioneering inventors and developers in the field of electrical and electronic telecommunications include Charles Wheatstone and Samuel Morse (inventors of the telegraph), Alexander Graham Bell (inventor of the telephone), Edwin Armstrong and Lee de Forest (inventors of radio), as well as Vladimir K. Zworykin, John Logie Baird and Philo Farnsworth (some of the inventors of television).

Earth station at the satellite communication facility in Raisting, Bavaria, Germany

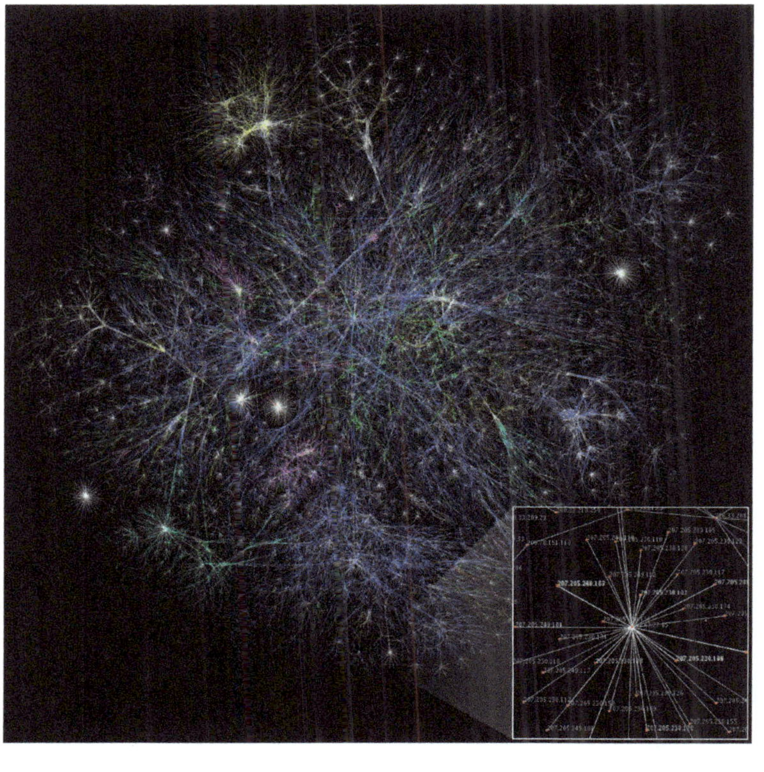

Visualization from the Opte Project of the various routes through a portion of the Internet

ETYMOLOGY

The word *telecommunication* is a compound of the Greek prefix *tele* (τηλε), meaning *distant*, *far off*, or *afar*, and the Latin *communicare*, meaning *to share*. Its modern use is adapted from the French, because its written use was recorded in 1904 by the French engineer and novelist Édouard Estaunié. *Communication* was first used as an English word in the late 14th century. It comes from Old French comunicacion (14c., Modern French communication), from Latin communicationem (nominative communicatio), noun of action from past participle stem of communicare "to share, divide out; communicate, impart, inform; join, unite, participate in," literally "to make common," from communis".

HISTORY

Beacons and Pigeons

In the Middle Ages, chains of beacons were commonly used on hilltops as a means of relaying a signal. Beacon chains suffered the drawback that they could only pass a single bit of information, so the meaning of the message such as "the enemy has been sighted" had to be agreed upon in advance. One notable instance of their use was during the Spanish Armada, when a beacon chain relayed a signal from Plymouth to London.

A replica of one of Chappe's semaphore towers

In 1792, Claude Chappe, a French engineer, built the first fixed visual telegraphy system (or semaphore line) between Lille and Paris. However semaphore suffered from the need for skilled operators and expensive towers at intervals of ten to thirty kilometres (six to nineteen miles). As a result of competition from the electrical telegraph, the last commercial line was abandoned in 1880. Homing pigeons have occasionally been used throughout history by different cultures. Pigeon post is thought to have Persians roots and was used by the Romans to aid their military. Frontinus said that Julius Caesar used pigeons as messengers in his conquest of Gaul. The Greeks also conveyed the names of the victors at the Olympic Games to various cities using homing pigeons. In the early 19th century, the Dutch government used the system in Java and Sumatra. And in 1849, Paul Julius Reuter started a pigeon service to fly stock prices between Aachen and Brussels, a service that operated for a year until the gap in the telegraph link was closed.

Telegraph and Telephone

Sir Charles Wheatstone and Sir William Fothergill Cooke invented the electric telegraph in 1837. Also, the first commercial electrical telegraph is purported to have been constructed by Wheatstone and Cooke and opened on 9 April 1839. Both inventors viewed their device as "an improvement to the [existing] electromagnetic telegraph" not as a new device.

Samuel Morse independently developed a version of the electrical telegraph that he unsuccessfully demonstrated on 2 September 1837. His code was an important advance over Wheatstone's signaling method. The first transatlantic telegraph cable was successfully completed on 27 July 1866, allowing transatlantic telecommunication for the first time.

The conventional telephone was invented independently by Alexander Bell and Elisha Gray in 1876. Antonio Meucci invented the first device that allowed the electrical transmission of voice over a line in 1849. However Meucci's device was of little practical value because it relied upon the electrophonic effect and thus required users to place the receiver in their mouth to "hear" what was being said. The first commercial telephone services were set-up in 1878 and 1879 on both sides of the Atlantic in the cities of New Haven and London.

Radio and Television

In 1832, James Lindsay gave a classroom demonstration of wireless telegraphy to his students. By 1854, he was able to demonstrate a transmission across the Firth of Tay from Dundee, Scotland to Woodhaven, a distance of two miles (3 km), using water as the transmission medium. In December 1901, Guglielmo Marconi established wireless communication between St. John's, Newfoundland (Canada) and Poldhu, Cornwall (England), earning him the 1909 Nobel Prize in physics (which he shared with Karl Braun). However small-scale radio communication had already been demonstrated in 1893 by Nikola Tesla in a presentation to the National Electric Light Association.

On 25 March 1925, John Logie Baird was able to demonstrate the transmission of moving pictures at the London department store Selfridges. Baird's device relied upon the Nipkow disk and thus became known as the mechanical television. It formed the basis of experimental broadcasts done by the British Broadcasting Corporation beginning 30 September 1929. However, for most of the twentieth century televisions depended upon the cathode ray tube invented by Karl Braun. The first version of such a television to show promise was produced by Philo Farnsworth and demonstrated to his family on 7 September 1927.

Computers and the Internet

On 11 September 1940, George Stibitz transmitted problems for his Complex Number Calculator in New York using a teletype, and received the computed results back at Dartmouth College in New Hampshire. This configuration of a centralized computer (mainframe) with remote dumb terminals remained popular well into the 1970s. However, already in the 1960s, researchers started to investigate packet switching, a technology that sends a message in portions to its destination asynchronously without passing it through a centralized mainframe. A four-node network emerged on 5 December 1969, constituting the beginnings of the ARPANET, which by 1981 had grown to 213 nodes. ARPANET eventually merged with other networks to form the Internet. While Internet development was a focus of the Internet Engineering Task Force (IETF) who published a series of Request for Comment documents, other networking advancement occurred in industrial laboratories, such as the local area network (LAN) developments of Ethernet (1983) and the token ring protocol (1984).

KEY CONCEPTS

Modern telecommunication is founded on a series of key concepts that experienced progressive development and refinement in a period of well over a century.

Basic Elements

Telecommunication technologies may primarily be divided into wired and wireless methods. Overall though, a basic telecommunication system consists of three main parts that are always present in some form or another:

- A transmitter that takes information and converts it to a signal.
- A transmission medium, also called the *physical channel* that carries the signal. An example of this is the "free space channel".
- A receiver that takes the signal from the channel and converts it back into usable information for the recipient.

For example, in a radio broadcasting station the station's large power amplifier is the transmitter; and the broadcasting antenna is the interface between the power amplifier and the "free space channel". The free space channel is the transmission medium; and the receiver's antenna is the interface between the free space channel and the receiver. Next, the radio receiver is the destination of the radio signal, and this is where it is converted from electricity to sound for people to listen to.

Sometimes, telecommunication systems are "duplex" (two-way systems) with a single box of electronics working as both the transmitter and a receiver, or a *transceiver*. For example, a cellular telephone is a transceiver. The transmission electronics and the receiver electronics within a transceiver are actually quite independent of each other. This can be readily explained by the fact that radio transmitters contain power amplifiers that operate with electrical powers measured in watts or kilowatts, but radio receivers deal with radio powers that are measured in the microwatts or nanowatts. Hence, transceivers have to be carefully designed and built to isolate their high-power circuitry and their low-power circuitry from each other, as to not cause interference.

Telecommunication over fixed lines is called point-to-point communication because it is between one transmitter and one receiver. Telecommunication through radio broadcasts is called broadcast communication because it is between one powerful transmitter and numerous low-power but sensitive radio receivers.

Telecommunications in which multiple transmitters and multiple receivers have been designed to cooperate and to share the same physical channel are called multiplex systems. The sharing of physical channels using multiplexing often gives very large reductions in costs. Multiplexed systems are laid out in telecommunication networks, and the multiplexed signals are switched at nodes through to the correct destination terminal receiver.

Analog Versus Digital Communications

Communications signals can be sent either by analog signals or digital signals. There are analog communication systems and digital communication systems. For an analog signal, the signal is varied continuously with respect to the information. In a digital signal, the information is encoded as a set of discrete values (for example, a set of ones and zeros). During the propagation and reception, the information contained in analog signals will inevitably be degraded by undesirable physical noise. (The output of a transmitter is noise-free for all practical purposes.) Commonly, the noise in a communication system can be expressed as adding or subtracting from the desirable signal in a completely random way. This form of noise is called additive noise, with the understanding that the noise can be negative or positive at different instants of time. Noise that is not additive noise is a much more difficult situation to describe or analyze, and these other kinds of noise will be omitted here.

On the other hand, unless the additive noise disturbance exceeds a certain threshold, the information contained in digital signals will remain intact. Their resistance to noise represents a key advantage of digital signals over analog signals.

Telecommunication Networks

A telecommunications network is a collection of transmitters, receivers, and communications channels that send messages to one another. Some digital communications networks contain one or more routers that work together to transmit information to the correct user. An analog communications network consists of one or more switches that establish a connection between two or more users. For both types of network, repeaters may be necessary to amplify or recreate the signal when it is being transmitted over long distances. This is to combat attenuation that can render the signal indistinguishable from the noise. Another advantage of digital systems over analog is that their output is easier to store in memory, i.e. two voltage states (high and low) are easier to store than a continuous range of states.

Communication Channels

The term "channel" has two different meanings. In one meaning, a channel is the physical medium that carries a signal between the transmitter and the receiver. Examples of this include the atmosphere for sound communications, glass optical fibers for some kinds of optical communications, coaxial cables for communications by way of the voltages and electric currents in them, and free space for communications using visible light, infrared waves, ultraviolet light, and radio waves. This last channel is called the "free space channel". The sending of radio waves from one place to another has nothing to do with the presence or absence of an atmosphere between the two. Radio waves travel through a perfect vacuum just as easily as they travel through air, fog, clouds, or any other kind of gas.

The other meaning of the term "channel" in telecommunications is seen in the phrase communications channel, which is a subdivision of a transmission medium so that it can be used to send multiple streams of information simultaneously. For example, one radio station can broadcast radio waves into free space at frequencies in the neighborhood of 94.5 MHz (megahertz) while another radio station can simultaneously broadcast radio waves at frequencies in the neighborhood of 96.1 MHz. Each radio station would transmit radio waves over a frequency bandwidth of about 180 kHz (kilohertz), centered at frequencies such as the above, which are called the "carrier frequencies". Each station in this example is separated from its adjacent stations by 200 kHz, and the difference between 200 kHz and 180 kHz (20 kHz) is an engineering allowance for the imperfections in the communication system.

In the example above, the "free space channel" has been divided into communications channels according to frequencies, and each channel is assigned a separate frequency bandwidth in which to broadcast radio waves. This system of dividing the medium into channels according to frequency is called "frequency-division multiplexing". Another term for the same concept is "wavelength-division multiplexing", which is more commonly used in optical communications when multiple transmitters share the same physical medium.

Another way of dividing a communications medium into channels is to allocate each sender a recurring segment of time (a "time slot", for example, 20 milliseconds out of each second), and to allow each sender to send messages only within its own time slot. This method of dividing the medium into communication channels is called "time-division multiplexing" (**TDM**), and is used in optical fiber communication. Some radio communication systems use TDM within an allocated FDM channel. Hence, these systems use a hybrid of TDM and FDM.

Modulation

The shaping of a signal to convey information is known as modulation. Modulation can be used to represent a digital message as an analog waveform. This is commonly called "keying" – a term derived from the older use of Morse Code in telecommunications – and several keying techniques exist (these include phase-shift keying, frequency-shift keying, and amplitude-shift keying). The "Bluetooth" system, for example, uses phase-shift keying to exchange information between various devices. In addition, there are combinations of phase-shift keying and amplitude-shift keying which is called (in the jargon of the field) "quadrature amplitude modulation" (QAM) that are used in high-capacity digital radio communication systems. Modulation can also be used to transmit the information of low-frequency analog signals at higher frequencies. This is helpful because low-frequency analog signals cannot be effectively transmitted over free space. Hence the information from a low-frequency analog signal must be impressed into a higher-frequency signal (known as the "carrier wave") before transmission. There are several different modulation schemes available to achieve this [two of the most basic being amplitude modulation (AM) and frequency modulation (FM)]. An example of this process is a disc jockey's voice being impressed into a 96 MHz carrier wave using frequency modulation (the voice would then be received on a radio as the channel "96 FM"). In addition, modulation has the advantage that it may use frequency division multiplexing (FDM).

SOCIETY

Telecommunication has a significant social, cultural and economic impact on modern society. In 2008, estimates placed the telecommunication industry's revenue at $4.7 trillion or just under 3 percent of the gross world product (official exchange rate). Several following sections discuss the impact of telecommunication on society.

Economic Impact

Microeconomics

On the microeconomic scale, companies have used telecommunications to help build global business empires. This is self-evident in the case of online retailer Amazon.com but, according to academic Edward Lenert, even the conventional retailer Walmart has benefited from better telecommunication infrastructure compared to its competitors. In cities throughout the world, home owners use their telephones to order and arrange a variety of home services ranging from pizza deliveries to electricians. Even relatively poor communities have been noted to use telecommunication to their advantage. In Bangladesh's Narshingdi district, isolated villagers use cellular phones to speak directly to wholesalers and arrange a better price for their goods. In Côte d'Ivoire, coffee growers share mobile phones to follow hourly variations in coffee prices and sell at the best price.

Macroeconomics

On the macroeconomic scale, Lars-Hendrik Röller and Leonard Waverman suggested a causal link between good telecommunication infrastructure and economic growth. Few dispute the existence of a correlation although some argue it is wrong to view the relationship as causal.

Because of the economic benefits of good telecommunication infrastructure, there is increasing worry about the inequitable access to telecommunication services amongst various countries of the world—this is known as the digital divide. A 2003 survey by the International Telecommunication Union (ITU) revealed that roughly a third of countries have fewer than one mobile subscription for every 20 people and one-third of countries have fewer than one land-line telephone subscription for every 20 people. In terms of Internet access, roughly half of all countries have fewer than one out of 20 people with Internet access. From this information, as well as educational data, the ITU was able to compile an index that measures the overall ability of citizens to access and use information and communication technologies. Using this measure, Sweden, Denmark and Iceland received the highest ranking while the African countries Nigeria, Burkina Faso and Mali received the lowest.

Social Impact

Telecommunication has played a significant role in social relationships. Nevertheless, devices like the telephone system were originally advertised with an emphasis on the practical dimensions of the device (such as the ability to conduct business or order home services) as opposed to the social dimensions. It was not until the late 1920s and 1930s that the social dimensions of the device became a prominent theme in telephone advertisements. New promotions started appealing to consumers' emotions, stressing the importance of social conversations and staying connected to family and friends.

Since then the role that telecommunications has played in social relations has become increasingly important. In recent years, the popularity of social networking sites has increased dramatically. These sites allow users to communicate with each other as well as post photographs, events and profiles for others to see. The profiles can list a person's age, interests, sexual preference and relationship status. In this way, these sites can play important role in everything from organising social engagements to courtship.

Prior to social networking sites, technologies like short message service (SMS) and the telephone also had a significant impact on social interactions. In 2000, market research group Ipsos MORI reported that 81% of 15- to 24-year-old SMS users in the United Kingdom had used the service to coordinate social arrangements and 42% to flirt.

Other impacts

News source preference of Americans in 2006.	
Local TV	59%
National TV	47%
Radio	44%
Local paper	38%
Internet	23%
National paper	12%
Survey permitted multiple answers	

In cultural terms, telecommunication has increased the public's ability to access music and film. With television, people can watch films they have not seen before in their own home without having to travel to the video store or cinema. With radio and the Internet, people can listen to music they have not heard before without having to travel to the music store.

Telecommunication has also transformed the way people receive their news. A 2006 survey (right table) of slightly more than 3,000 Americans by the non-profit Pew Internet and American Life Project in the United States the majority specified television or radio over newspapers.

Telecommunication has had an equally significant impact on advertising. TNS Media Intelligence reported that in 2007, 58% of advertising expenditure in the United States was spent on media that depend upon telecommunication.

Advertising expenditures in US in 2007		
Medium		**Spending**
Internet	7.6%	$11.31 billion
Radio	7.2%	$10.69 billion
Cable TV	12.1%	$18.02 billion
Syndicated TV	2.8%	$4.17 billion
Spot TV	11.3%	$16.82 billion
Network TV	17.1%	$25.42 billion
Newspaper	18.9%	$28.22 billion
Magazine	20.4%	$30.33 billion
Outdoor	2.7%	$4.02 billion
Total	100%	$149 billion

GOVERNMENT

Many countries have enacted legislation which conforms to the *International Telecommunication Regulations* established by the International Telecommunication Union (ITU), which is the "leading UN agency for information and communication technology issues." In 1947, at the Atlantic City Conference, the ITU decided to "afford international protection to all frequencies registered in a new international frequency list and used in conformity with the Radio Regulation." According to the ITU's *Radio Regulations* adopted in Atlantic City, all frequencies referenced in the *International Frequency Registration Board*, examined by the board and registered on the *International Frequency List* "shall have the right to international protection from harmful interference."

From a global perspective, there have been political debates and legislation regarding the management of telecommunication and broadcasting. The history of broadcasting discusses some debates in relation to balancing conventional communication such as printing and telecommunication such as radio broadcasting. The onset of World War II brought on the first explosion of international broadcasting propaganda. Countries, their governments, insurgents, terrorists, and militiamen have all used telecommunication and broadcasting techniques to promote propaganda. Patriotic propaganda for political movements and colonization started the mid-1930s. In 1936, the BBC broadcast propaganda to the Arab World to partly counter similar broadcasts from Italy, which also had colonial interests in North Africa.

Modern insurgents, such as those in the latest Iraq war, often use intimidating telephone calls, SMSs and the distribution of sophisticated videos of an attack on coalition troops within hours of the operation. "The Sunni insurgents even have their own television station, Al-Zawraa, which while banned by the Iraqi government, still broadcasts from Erbil, Iraqi Kurdistan, even as coalition pressure has forced it to switch satellite hosts several times."

On 10 November 2014, President Obama recommended the Federal Communications Commission reclassify broadband Internet service as a telecommunications service in order to preserve net neutrality.

MODERN MEDIA

Worldwide Equipment Sales

According to data collected by Gartner and Ars Technica sales of main consumer's telecommunication equipment worldwide in millions of units was:

Equipment / year	1975	1980	1985	1990	1994	1996	1998	2000	2002	2004	2006	2008
Computers	0	1	8	20	40	75	100	135	130	175	230	280
Cell phones	N/A	N/A	N/A	N/A	N/A	N/A	180	400	420	660	830	1000

Telephone

In a telephone network, the caller is connected to the person they want to talk to by switches at various telephone exchanges. The switches form an electrical connection between the two users and the setting of these switches is determined electronically when the caller dials the number. Once the connection is made, the caller's voice is transformed to an electrical signal using a small microphone in the caller's handset. This electrical signal is then sent through the network to the user at the other end where it is transformed back into sound by a small speaker in that person's handset.

Optical fiber provides cheaper bandwidth for long distance communication.

The landline telephones in most residential homes are analog—that is, the speaker's voice directly determines the signal's voltage. Although short-distance calls may be handled from end-to-end as analog signals, increasingly telephone service providers are transparently converting the signals to digital signals for transmission. The advantage of this is that digitized voice data can travel side-by-side with data from the Internet and can be perfectly reproduced in long distance communication (as opposed to analog signals that are inevitably impacted by noise).

Mobile phones have had a significant impact on telephone networks. Mobile phone subscriptions now outnumber fixed-line subscriptions in many markets. Sales of mobile phones in 2005 totalled 816.6 million with that figure being almost equally shared amongst the markets of Asia/Pacific (204 m), Western Europe (164 m), CEMEA (Central Europe, the Middle East and Africa) (153.5 m), North America (148 m) and Latin America (102 m). In terms of new subscriptions over the five years from 1999, Africa has outpaced other markets with 58.2% growth Increasingly these phones are being serviced by systems where the voice content is transmitted digitally such as GSM or W-CDMA with many markets choosing to deprecate analog systems such as AMPS.

There have also been dramatic changes in telephone communication behind the scenes. Starting with the operation of TAT-8 in 1988, the 1990s saw the widespread adoption of systems based on optical fibers. The benefit of communicating with optic fibers is that they offer a drastic increase in data capacity. TAT-8 itself was able to carry 10 times as many telephone calls as the last copper cable laid at that time and today's optic fibre cables are able to carry 25 times as many telephone calls as TAT-8. This increase in data capacity is due to several factors: First, optic fibres are physically much smaller than competing technologies. Second, they do not suffer from crosstalk which means several hundred of them can be easily bundled together in a single cable. Lastly, improvements in multiplexing have led to an exponential growth in the data capacity of a single fibre.

Assisting communication across many modern optic fibre networks is a protocol known as Asynchronous Transfer Mode (ATM). The ATM protocol allows for the side-by-side data transmission mentioned in the second paragraph. It is suitable for public telephone networks because it establishes a pathway for data through the network and associates a traffic contract with that pathway. The traffic contract is essentially an agreement between the client and the network about how the network is to handle the data; if the network cannot meet the conditions of the traffic contract it does not accept the connection. This is important because telephone calls can negotiate a contract so as to guarantee themselves a constant bit rate, something that will ensure a caller's voice is not delayed in parts or cut off completely. There are competitors to ATM, such as Multiprotocol Label Switching (MPLS), that perform a similar task and are expected to supplant ATM in the future.

Radio and Television

In a broadcast system, the central high-powered broadcast tower transmits a high-frequency electromagnetic wave to numerous low-powered receivers. The high-frequency wave sent by the tower is modulated with a signal containing visual or audio information. The receiver is then tuned so as to pick up the high-frequency wave and a demodulator is used to retrieve the signal containing the visual or audio information. The broadcast signal can be either analog (signal is varied continuously with respect to the information) or digital (information is encoded as a set of discrete values).

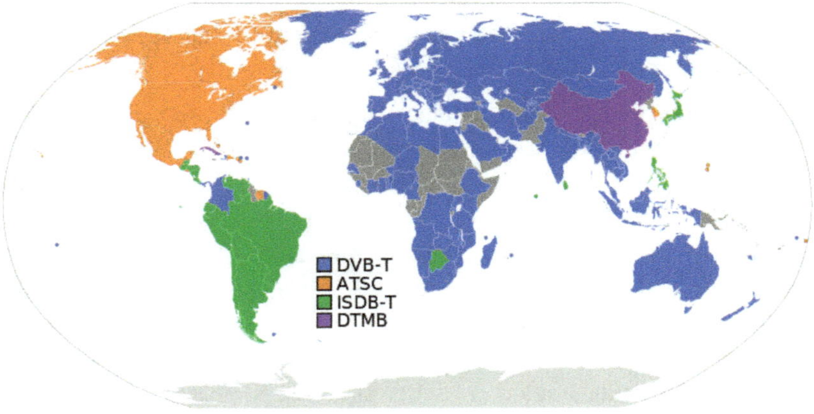

Digital television standards and their adoption worldwide

The broadcast media industry is at a critical turning point in its development, with many countries moving from analog to digital broadcasts. This move is made possible by the production of cheaper, faster and more capable integrated circuits. The chief advantage of digital broadcasts is that they prevent a number of complaints common to traditional analog broadcasts. For television, this includes the elimination of problems such as snowy pictures, ghosting and other distortion. These occur because of the nature of analog transmission, which means that perturbations due to noise will be evident in the final output. Digital transmission overcomes this problem because digital signals are reduced to discrete values upon reception and hence small perturbations do not affect the final output. In a simplified example, if a binary message 1011 was transmitted with signal amplitudes [1.0 0.0 1.0 1.0] and received with signal amplitudes [0.9 0.2 1.1 0.9] it would still decode to the binary message 1011 — a perfect reproduction of what was sent. From this example, a problem with digital transmissions can also be seen in that if the noise is great enough it can significantly alter the decoded message. Using forward error correction a receiver can correct a handful of bit errors in the resulting message but too much noise will lead to incomprehensible output and hence a breakdown of the transmission.

In digital television broadcasting, there are three competing standards that are likely to be adopted worldwide. These are the ATSC, DVB and ISDB standards; the adoption of these standards thus far is presented in the captioned map. All three standards use MPEG-2 for video compression. ATSC uses Dolby Digital AC-3 for audio compression, ISDB uses Advanced Audio Coding (MPEG-2 Part 7) and DVB has no standard for audio compression but typically uses MPEG-1 Part 3 Layer 2. The choice of modulation also varies between the schemes. In digital audio broadcasting, standards are much more unified with practically all countries choosing to adopt the Digital Audio Broadcasting standard (also known as the Eureka 147 standard). The exception is the United States which has chosen to adopt HD Radio. HD Radio, unlike Eureka 147, is based upon a transmission method known as in-band on-channel transmission that allows digital information to "piggyback" on normal AM or FM analog transmissions.

However, despite the pending switch to digital, analog television remains being transmitted in most countries. An exception is the United States that ended analog television transmission (by all but the very low-power TV stations) on 12 June 2009 after twice delaying the switchover deadline,Kenya also ended analog television transmission in December 2014 after multiple delays. For analog television, there are three standards in use for broadcasting color TV (see a map on adoption here). These are known as PAL (German designed), NTSC (North American designed), and SECAM (French designed). (It is important to understand that these are the ways of sending color TV, and they do not have anything to do with the standards for black & white TV, which also vary from country to country.) For analog radio, the switch to digital radio is made more difficult by the fact that analog receivers are sold at a small fraction of the price of digital receivers. The choice of modulation for analog radio is typically between amplitude (**AM**) or frequency modulation (**FM**). To achieve stereo playback, an amplitude modulated subcarrier is used for stereo FM.

Internet

The Internet is a worldwide network of computers and computer networks that communicate with each other using the Internet Protocol. Any computer on the Internet has a unique IP address that can be used by other computers to route information to it. Hence, any computer on the Internet can send a message to any other computer using its IP address. These messages carry with them the originating computer's IP address allowing for two-way communication. The Internet is thus an exchange of messages between computers.

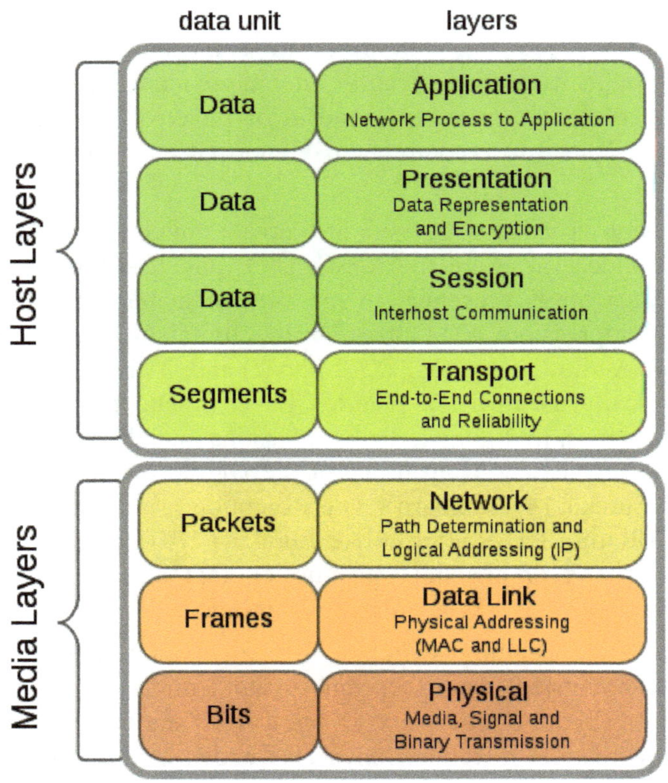

The OSI reference model

It is estimated that the 51% of the information flowing through two-way telecommunications networks in the year 2000 were flowing through the Internet (most of the rest (42%) through the landline telephone). By the year 2007 the Internet clearly dominated and captured 97% of all the information in telecommunication networks (most of the rest (2%) through mobile phones). As of 2008, an estimated 21.9% of the world population has access to the Internet with the highest access rates (measured as a percentage of the population) in North America (73.6%), Oceania/Australia (59.5%) and Europe (48.1%). In terms of broadband access, Iceland (26.7%), South Korea (25.4%) and the Netherlands (25.3%) led the world.

The Internet works in part because of protocols that govern how the computers and routers communicate with each other. The nature of computer network communication lends itself to a layered approach where individual protocols in the protocol stack run more-or-less independently of other protocols. This allows lower-level protocols to be customized for the network situation while not changing the way higher-level protocols operate. A practical example of why this is important is because it allows an Internet browser to run the same code regardless of whether the computer it is running on is connected to the Internet through an Ethernet or Wi-Fi connection. Protocols are often talked about in terms of their place in the OSI reference model (pictured on the right), which emerged in 1983 as the first step in an unsuccessful attempt to build a universally adopted networking protocol suite.

For the Internet, the physical medium and data link protocol can vary several times as packets traverse the globe. This is because the Internet places no constraints on what physical medium or data link protocol is used. This leads to the adoption of media and protocols that best suit the local network situation. In practice, most intercontinental communication will use the Asynchronous Transfer Mode (ATM) protocol (or a modern equivalent) on top of optic fiber. This is because for most intercontinental communication the Internet shares the same infrastructure as the public switched telephone network.

At the network layer, things become standardized with the Internet Protocol (IP) being adopted for logical addressing. For the World Wide Web, these "IP addresses" are derived from the human readable form using the Domain Name System (e.g. 72.14.207.99 is derived from www.google.com). At the moment, the most widely used version of the Internet Protocol is version four but a move to version six is imminent.

At the transport layer, most communication adopts either the Transmission Control Protocol (TCP) or the User Datagram Protocol (UDP). TCP is used when it is essential every message sent is received by the other computer whereas UDP is used when it is merely desirable. With TCP, packets are retransmitted if they are lost and placed in order before they are presented to higher layers. With UDP, packets are not ordered or retransmitted if lost. Both TCP and UDP packets carry port numbers with them to specify what application or process the packet should be handled by. Because certain application-level protocols use certain ports, network administrators can manipulate traffic to suit particular requirements. Examples are to restrict Internet access by blocking the traffic destined for a particular port or to affect the performance of certain applications by assigning priority.

Above the transport layer, there are certain protocols that are sometimes used and loosely fit in the session and presentation layers, most notably the Secure Sockets Layer (SSL) and Transport Layer Security (TLS) protocols. These protocols ensure that data transferred between two parties remains completely confidential. Finally, at the application layer, are many of the protocols Internet users would be familiar with such as HTTP (web browsing), POP3 (e-mail), FTP (file transfer), IRC (Internet chat), BitTorrent (file sharing) and XMPP (instant messaging).

Voice over Internet Protocol (VoIP) allows data packets to be used for synchronous voice communications. The data packets are marked as voice type packets and can be prioritized by the network administrators so that the real-time, synchronous conversation is less subject to contention with other types of data traffic which can be delayed (i.e.

file transfer or email) or buffered in advance (i.e. audio and video) without detriment. That prioritization is fine when the network has sufficient capacity for all the VoIP calls taking place at the same time and the network is enabled for prioritization i.e. a private corporate style network, but the Internet is not generally managed in this way and so there can be a big difference in the quality of VoIP calls over a private network and over the public Internet.

Local Area Networks and Wide Area Networks

Despite the growth of the Internet, the characteristics of local area networks (LANs)--computer networks that do not extend beyond a few kilometers—remain distinct. This is because networks on this scale do not require all the features associated with larger networks and are often more cost-effective and efficient without them. When they are not connected with the Internet, they also have the advantages of privacy and security. However, purposefully lacking a direct connection to the Internet does not provide assured protection from hackers, military forces, or economic powers. These threats exist if there are any methods for connecting remotely to the LAN.

Wide area networks (WANs) are private computer networks that may extend for thousands of kilometers. Once again, some of their advantages include privacy and security. Prime users of private LANs and WANs include armed forces and intelligence agencies that must keep their information secure and secret.

In the mid-1980s, several sets of communication protocols emerged to fill the gaps between the data-link layer and the application layer of the OSI reference model. These included Appletalk, IPX, and NetBIOS with the dominant protocol set during the early 1990s being IPX due to its popularity with MS-DOS users. TCP/IP existed at this point, but it was typically only used by large government and research facilities.

As the Internet grew in popularity and its traffic was required to be routed into private networks, the TCP/IP protocols replaced existing local area network technologies. Additional technologies, such as DHCP, allowed TCP/IP-based computers to self-configure in the network. Such functions also existed in the AppleTalk/ IPX/ NetBIOS protocol sets.

Whereas Asynchronous Transfer Mode (ATM) or Multiprotocol Label Switching (MPLS) are typical data-link protocols for larger networks such as WANs; Ethernet and Token Ring are typical data-link protocols for LANs. These protocols differ from the former protocols in that they are simpler, e.g., they omit features such as quality of service guarantees, and offer collision prevention. Both of these differences allow for more economical systems.

Despite the modest popularity of IBM Token Ring in the 1980s and 1990s, virtually all LANs now use either wired or wireless Ethernet facilities. At the physical layer, most wired Ethernet implementations use copper twisted-pair cables (including the common 10BASE-T networks). However, some early implementations used heavier coaxial cables and some recent implementations (especially high-speed ones) use optical fibers. When optic fibers are used, the distinction must be made between multimode fibers and single-mode fibers. Multimode fibers can be thought of as thicker optical fibers that are cheaper to manufacture devices for, but that suffers from less usable bandwidth and worse attenuation – implying poorer long-distance performance.

TRANSMISSION CAPACITY

The effective capacity to exchange information worldwide through two-way telecommunication networks grew from 281 petabytes of (optimally compressed) information in 1986, to 471 petabytes in 1993, to 2.2 (optimally compressed) exabytes in 2000, and to 65 (optimally compressed) exabytes in 2007. This is the informational equivalent of two newspaper pages per person per day in 1986, and six entire newspapers per person per day by 2007. Given this growth, telecommunications play an increasingly important role in the world economy and the global telecommunications industry was about a $4.7 trillion sector in 2012. The service revenue of the global telecommunications industry was estimated to be $1.5 trillion in 2010, corresponding to 2.4% of the world's gross domestic product (GDP).

REFERENCES

1. "Article 1.3" (PDF), ITU Radio Regulations, International Telecommunication Union, 2012
2. Constitution and Convention of the International Telecommunication Union, Annex (Geneva, 1992)
3. "Definition of telecommunication". Yahoo. Retrieved 28 February 2013.
4. "Telecommunication". Collins English Dictionary. Retrieved 28 February 2013.
5. "Telecommunication". Vocabulary.com. Retrieved 28 February 2013.
6. "Telecommunication". Merriam-Webster Dictionary. Retrieved 28 February 2013.
7. "Telecommunication". Oxford Dictionaries. Oxford University Press. Retrieved 28 February 2013.
8. "Telecommunication". Dictionary.com. Retrieved 28 February 2013.
9. http://www.etymonline.com/index.php?term=tele-&allowed_in_frame=0
10. Jean-Marie Dilhac, From tele-communicare to Telecommunications, 2004.
11. Telecommunication, tele- and communication, New Oxford American Dictionary (2nd edition), 2005.
12. http://www.etymonline.com/index.php?term=communication&allowed_in_frame=0
13. David Ross, The Spanish Armada, Britain Express, accessed October 2007.
14. Les Télégraphes Chappe, Cédrick Chatenet, l'Ecole Centrale de Lyon, 2003.
15. CCIT/ITU-T 50 Years of Excellence, International Telecommunication Union, 2006.
16. Levi, Wendell (1977). The Pigeon. Sumter, S.C.: Levi Publishing Co, Inc. ISBN 0853900132.
17. Blechman, Andrew (2007). Pigeons-The fascinating saga of the world's most

revered and reviled bird. St Lucia, Queensland: University of Queensland Press. ISBN 9780702236419.

18. "Chronology: Reuters, from pigeons to multimedia merger" (Web article). Reuters. Retrieved 2008-02-21.
19. William Brockedone. "Cooke and Wheatstone and the Invention of the Electric Telegraph". Republished by The Museum of Science and Technology (Ottawa).
20. The Electromagnetic Telegraph, J. B. Calvert, 19 May 2004.
21. The Atlantic Cable, Bern Dibner, Burndy Library Inc., 1959
22. Elisha Gray, Oberlin College Archives, Electronic Oberlin Group, 2006.
23. Antonio Santi Giuseppe Meucci, Eugenii Katz. (Retrieved May, 2006 from http://chem.ch.huji.ac.il/~eugeniik/history/meucci.html)
24. Connected Earth: The telephone, BT, 2006.
25. History of AT&T, AT&T, 2006.
26. James Bowman Lindsay, Macdonald Black, Dundee City Council, 1999.
27. Tesla Biography, Ljubo Vujovic, Tesla Memorial Society of New York, 1998.
28. Tesla's Radio Controlled Boat, Twenty First Century Books, 2007.
29. The Pioneers, MZTV Museum of Television, 2006.
30. Philo Farnsworth, Neil Postman, TIME Magazine, 29 March 1999
31. George Stlibetz, Kerry Redshaw, 1996.
32. Hafner, Katie (1998). Where Wizards Stay Up Late: The Origins Of The Internet. Simon & Schuster. ISBN 0-684-83267-4.
33. Haykin, Simon (2001). Communication Systems (4th ed.). John Wiley & Sons. pp. 1–3. ISBN 0-471-17869-1.
34. Ambardar, Ashok (1999). Analog and Digital Signal Processing (2nd ed.). Brooks/Cole Publishing Company. pp. 1–2. ISBN 0-534-95409-X.
35. ATIS Telecom Glossary 2000, ATIS Committee T1A1 Performance and Signal Processing (approved by the American National Standards Institute), 28 February 2001.
36. Haykin, pp 344–403.
37. Bluetooth Specification Version 2.0 + EDR (p 27), Bluetooth, 2004.
38. Haykin, pp 88–126.
39. Worldwide Telecommunications Industry Revenues, Internet Engineering Task Force, June 2010.
40. Lenert, Edward (December 1998). "A Communication Theory Perspective on Telecommunications Policy". Journal of Communication. 48 (4): 3–23. doi:10.1111/j.1460-2466.1998.tb02767.x.
41. Mireille Samaan (April 2003). "The Effect of Income Inequality on Mobile Phone Penetration". Boston University Honors thesis. Archived from the original

(PDF) on 14 February 2007. Retrieved 8 June 2007.

42. Röller, Lars-Hendrik; Leonard Waverman (2001). "Telecommunications Infrastructure and Economic Development: A Simultaneous Approach". American Economic Review. 91 (4): 909–923. doi:10.1257/aer.91.4.909. ISSN 0002-8282.
43. Qiang, Rossotto, 2009, Economic Impact of Broadband, http://siteresources.worldbank.org/EXTIC4D/Resources/IC4D_Broadband_35_50.pdf
44. Riaz, Ali (1997). "The role of telecommunications in economic growth: proposal for an alternative framework of analysis". Media, Culture & Society. 19 (4): 557–583. doi:10.1177/016344397019004004.
45. "Digital Access Index (DAI)". itu.int. Retrieved 6 March 2008.
46. World Telecommunication Development Report 2003, International Telecommunication Union, 2003.
47. Fischer, Claude S. "'Touch Someone': The Telephone Industry Discovers Sociability." Technology and Culture 29.1 (January 1988): 32–61. JSTOR. Web. 4 October 2009.
48. "How do you know your love is real? Check Facebook". CNN. 4 April 2008.
49. I Just Text To Say I Love You, Ipsos MORI, September 2005.
50. "Online News: For many home broadband users, the internet is a primary news source" (PDF). Pew Internet Project. 22 March 2006.
51. "100 Leading National Advertisers" (PDF). Advertising Age. 23 June 2008. Retrieved 21 June 2009.
52. International Telecommunication Union : About ITU. ITU. Accessed 21 July 2009. (PDF of regulation)
53. Codding, George A. Jr. "Jamming and the Protection of Frequency Assignments". The American Journal of International Law, Vol. 49, No. 3 (July , 1955), Published by: American Society of International Law. pp. 384–388. Republished by JSTOR.org The American Journal of International Law". Accessed 21 July 2009.
54. Wood, James & Science Museum (Great Britain) "History of international broadcasting". IET 1994, Volume 1, p.2 of 258 ISBN 0-86341-302-1, ISBN 978-0-86341-302-5. Republished by Googlebooks. Accessed 21 July 2009.
55. Garfield, Andrew. "The U.S. Counter-propaganda Failure in Iraq", FALL 2007, The Middle East Quarterly, Volume XIV: Number 4, Accessed 21 July 2009.
56. Wyatt, Edward (10 November 2014). "Obama Asks F.C.C. to Adopt Tough Net Neutrality Rules". New York Times. Retrieved 15 November 2014.
57. NYT Editorial Board (14 November 2014). "Why the F.C.C. Should Heed President Obama on Internet Regulation". New York Times. Retrieved 15 November 2014.
58. Computer sales review, guardian.co.uk, 2009.
59. Mobile phone sales data, palminfocenter.com, 2009.
60. PC early history, arstechnica.com, 2005.

61. Gartner Says Top Six Vendors Drive Worldwide Mobile Phone Sales to 21% Growth in 2005, Gartner Group, 28 February 2006.
62. Africa Calling, Victor and Irene Mbarika, IEEE Spectrum, May 2006.
63. Ten Years of GSM in Australia, Australia Telecommunications Association, 2003.
64. Milestones in AT&T History, AT&T Knowledge Ventures, 2006.
65. Optical fibre waveguide, Saleem Bhatti, 1995.
66. Fundamentals of DWDM Technology, CISCO Systems, 2006.
67. Report: DWDM No Match for Sonet, Mary Jander, Light Reading, 2006.
68. Stallings, William (2004). Data and Computer Communications (7th edition (intl) ed.). Pearson Prentice Hall. pp. 337–366. ISBN 0-13-183311-1.
69. MPLS is the future, but ATM hangs on, John Dix, Network World, 2002
70. Lazar, Irwin (22 February 2011). "The WAN Road Ahead: Ethernet or Bust?". Telecom Industry Updates. Retrieved 22 February 2011.
71. How Radio Works, HowStuffWorks.com, 2006.
72. Digital Television in Australia, Digital Television News Australia, 2001.
73. Stallings, William (2004). Data and Computer Communications (7th edition (intl) ed.). Pearson Prentice Hall. ISBN 0-13-183311-1.
74. HDV Technology Handbook, Sony, 2004.
75. Audio, Digital Video Broadcasting Project, 2003.
76. Status of DAB (USA), World DAB Forum, March 2005.
77. Brian Stelter (13 June 2009). "Changeover to Digital TV Off to a Smooth Start". New York Times.
78. GE 72664 Portable AM/FM Radio, Amazon.com, June 2006.
79. DAB Products, World DAB Forum, 2006.
80. Robert E. Kahn and Vinton G. Cerf, What Is The Internet (And What Makes It Work), December 1999. (specifically see footnote xv)
81. How Internet Infrastructure Works, HowStuffWorks.com, 2007.
82. "The World's Technological Capacity to Store, Communicate, and Compute Information", Martin Hilbert and Priscila López (2011), Science, 332(6025), 60–65; free access to the study through here: martinhilbert.net/WorldInfoCapacity.html
83. World Internet Users and Population Stats, internetworldstats.com, 19 March 2007.
84. OECD Broadband Statistics, Organisation for Economic Co-operation and Development, December 2005.
85. History of the OSI Reference Model, The TCP/IP Guide v3.0, Charles M. Kozierok, 2005.
86. Introduction to IPv6, Microsoft Corporation, February 2006.
87. Stallings, pp 683–702.
88. T. Dierks and C. Allen, The TLS Protocol Version 1.0, RFC 2246, 1999.

89. Multimedia, Crucible (2011-05-07). "VoIP, Voice over Internet Protocol and Internet telephone calls".
90. Martin, Michael (2000). Understanding the Network (The Networker's Guide to AppleTalk, IPX, and NetBIOS), SAMS Publishing, ISBN 0-7357-0977-7.
91. Ralph Droms, Resources for DHCP, November 2003.
92. Stallings, pp. 500–526.
93. Stallings, pp 514–516.
94. Fiber Optic Cable Tutorial, Arc Electronics. Retrieved June 2007.
95. "video animation The Economist".
96. Introduction to the Telecommunications Industry, Internet Engineering Task Force, June 2012.

CHAPTER 8

Instrumentation System

Chapter Contents

- **Introduction**
- **History and Development**
- **Applications**
- **Measurement Parameters**
- **Instrumentation Engineering**
- **Impact of Modern Development**

The basic need of instrumentation in a process is to get the best and most amount of information so as to successfully complete the process. When referring to the completion of the project with reference to instrumentation, it basically means maximum efficiency with minimum production expense and desired output quality.

The information that is achieved from these processes may be very simple and may mostly involve a direct measurement method. But as the process becomes more complex, direct measurement may seem to be impracticable and so indirect methods must be used for measurements. These methods involve a derived relationship between the measured quantity and the result that is needed.

Most of the indirect methods involve electrical techniques as they have high speed and also simple processing methods. The output from such methods is easier to link to computers.

INTRODUCTION

Instrumentation is a collective term for measuring instruments used for indicating, measuring and recording physical quantities.

The term instrumentation may refer to something as simple as direct reading thermometers or, when using many sensors, may become part of a complex Industrial control system in such as manufacturing industry, vehicles and transportation. Instrumentation can be found in the household as well; a smoke detector or a heating thermostat are examples.

HISTORY AND DEVELOPMENT

A local instrumentation panel on a steam turbine.

The history of instrumentation can be divide into several phases.

Pre-Industrial

Elements of industrial instrumentation have long histories. Scales for comparing weights and simple pointers to indicate position are ancient technologies. Some of the earliest measurements were of time. One of the oldest water clocks was found in the tomb of the Egyptian pharaoh Amenhotep I, buried around 1500 BCE. Improvements were incorporated in the clocks. By 270 BCE they had the rudiments of an automatic control system device.

In 1663 Christopher Wren presented the Royal Society with a design for a "weather clock". A drawing shows meteorological sensors moving pens over paper driven by clockwork. Such devices did not become standard in meteorology for two centuries. The concept has remained virtually unchanged as evidenced by pneumatic chart recorders, where a pressurized bellows displaces a pen. Integrating sensors, displays, recorders and controls was uncommon until the industrial revolution, limited by both need and practicality.

Early Industrial

Electronics enabled wiring to replace pipes. A transmitter is a device that produces an output signal, often in the form of a 4–20 mA electrical current signal, although many other options using voltage, frequency, pressure, or ethernet are possible. The transistor was commercialized by the mid-1950s.

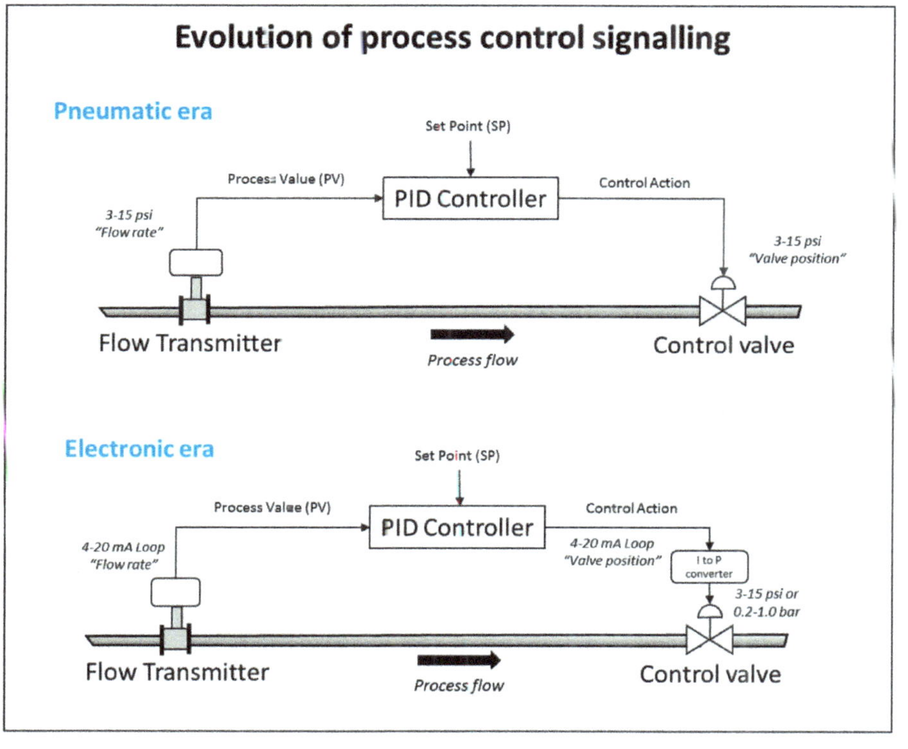

The evolution of analogue control loop signaling from the pneumatic era to the electronic era.

Instruments attached to a control system provided signals used to operate solenoids, valves, regulators, circuit breakers, relays and other devices. Such devices could control a desired output variable, and provide either remote or automated control capabilities.

Each instrument company introduced their own standard instrumentation signal, causing confusion until the 4-20 mA range was used as the standard electronic instrument signal for transmitters and valves. This signal was eventually standardized as ANSI/ISA S50, "Compatibility of Analog Signals for Electronic Industrial Process Instruments", in the 1970s. The transformation of instrumentation from mechanical pneumatic transmitters, controllers, and valves to electronic instruments reduced maintenance costs as electronic instruments were more dependable than mechanical instruments. This also increased efficiency and production due to their increase in accuracy. Pneumatics enjoyed some advantages, being favored in corrosive and explosive atmospheres.

Automatic Process Control

In the early years of process control, process indicators and control elements such as valves were monitored by an operator that walked around the unit adjusting the valves to obtain the desired temperatures, pressures, and flows. As technology evolved pneumatic controllers were invented and mounted in the field that monitored the process and controlled the valves. This reduced the amount of time process operators were needed to monitor the process. Later years the actual controllers were moved to a central room and signals were sent into the control room to monitor the process and outputs signals were sent to the final control element such as a valve to adjust the process as needed. These controllers and indicators were mounted on a wall called a control board. The operators stood in front of this board walking back and forth monitoring the process indicators. This again reduced the number and amount of time process operators were needed to walk around the units. The most standard pneumatic signal level used during these years was 3-15 psig.

Large Integrated Computer-Based Systems

Process control of large industrial plants has evolved through many stages. Initially, control would be from panels local to the process plant. However this required a large manpower resource to attend to these dispersed panels, and there was no overall view of the process. The next logical development was the transmission of all plant measurements to a permanently-manned central control room. Effectively this was the centralization of all the localized panels, with the advantages of lower manning levels and easier overview of the process. Often the controllers were behind the control room panels, and all automatic and manual control outputs were transmitted back to plant.

Pneumatic "Three term" pneumatic PID controller, widely used before electronics became reliable and cheaper

A pre-DCS/SCADA era central control room. Whilst the controls are centralized in one place, they are still discrete and not integrated into one system.

A DCS control room where plant information and controls are displayed on computer graphics screens. The operators are seated and can view and control any part of the process from their screens, whilst retaining a plant overview.

However, whilst providing a central control focus, this arrangement was inflexible as each control loop had its own controller hardware, and continual operator movement within the control room was required to view different parts of the process. With coming of electronic processors and graphic displays it became possible to replace these discrete controllers with computer-based algorithms, hosted on a network of input/output racks with their own control processors. These could be distributed around plant, and communicate with the graphic display in the control room or rooms. The distributed control concept was born.

The introduction of DCSs and SCADA allowed easy interconnection and re-configuration of plant controls such as cascaded loops and interlocks, and easy interfacing with other production computer systems. It enabled sophisticated alarm handling, introduced automatic event logging, removed the need for physical records such as chart recorders, allowed the control racks to be networked and thereby located locally to plant to reduce cabling runs, and provided high level overviews of plant status and production levels.

APPLICATIONS

In some cases the sensor is a very minor element of the mechanism. Digital cameras and wristwatches might technically meet the loose definition of instrumentation because they record and/or display sensed information. Under most circumstances neither would be called instrumentation, but when used to measure the elapsed time of a race and to document the winner at the finish line, both would be called instrumentation.

Household

A very simple example of an instrumentation system is a mechanical thermostat, used to control a household furnace and thus to control room temperature. A typical unit senses temperature with a bi-metallic strip. It displays temperature by a needle on the free end of the strip. It activates the furnace by a mercury switch. As the switch is rotated by the strip, the mercury makes physical (and thus electrical) contact between electrodes.

Another example of an instrumentation system is a home security system. Such a system consists of sensors (motion detection, switches to detect door openings), simple algorithms to detect intrusion, local control (arm/disarm) and remote monitoring of the system so that the police can be summoned. Communication is an inherent part of the design.

Kitchen appliances use sensors for control.

- A refrigerator maintains a constant temperature by measuring the internal temperature.
- A microwave oven sometimes cooks via a heat-sense-heat-sense cycle until sensing done.
- An automatic ice machine makes ice until a limit switch is thrown.
- Pop-up bread toasters can operate by time or by heat measurements.
- Some ovens use a temperature probe to cook until a target internal food temperature is reached.

Automotive

Modern automobiles have complex instrumentation. In addition to displays of engine rotational speed and vehicle linear speed, there are also displays of battery voltage and current, fluid levels, fluid temperatures, distance traveled and feedbacks of various controls (turn signals, parking brake, headlights, transmission position). Cautions may be displayed for special problems (fuel low, check engine, tire pressure low, door ajar, seat belt unfastened). Problems are recorded so they can be reported to diagnostic equipment. Navigation systems can provide voice commands to reach a destination. Automotive instrumentation must be cheap and reliable over long periods in harsh environments. There may be independent airbag systems which contain sensors, logic and actuators. Anti-skid braking systems use sensors to control the brakes, while cruise control affects throttle position. A wide variety of services can be provided via communication links as the OnStar system. Autonomous cars (with exotic instrumentation) have been demonstrated.

Aircraft

Early aircraft had a few sensors. "Steam gauges" converted air pressures into needle deflections that could be interpreted as altitude and airspeed. A magnetic compass provided a sense of direction. The displays to the pilot were as critical as the measurements.

A modern aircraft has a far more sophisticated suite of sensors and displays, which are embedded into avionics systems. The aircraft may contain inertial navigation systems, global positioning systems, weather radar, autopilots, and aircraft stabilization systems. Redundant sensors are used for reliability. A subset of the information may be transferred to a crash recorder to aid mishap investigations. Modern pilot displays now include computer displays including head-up displays.

Air traffic control radar is distributed instrumentation system. The ground portion transmits an electromagnetic pulse and receives an echo (at least). Aircraft carry transponders that transmit codes on reception of the pulse. The system displays aircraft map location, an identifier and optionally altitude. The map location is based on sensed antenna direction and sensed time delay. The other information is embedded in the transponder transmission.

Laboratory instrumentation

Among the Possible uses of the term is a collection of laboratory test equipment controlled by a computer through an IEEE-488 bus (also known as GPIB for General Purpose Instrument Bus or HPIB for Hewlitt Packard Instrument Bus). Laboratory equipment is available to measure many electrical and chemical quantities. Such a collection of equipment might be used to automate the testing of drinking water for pollutants.

MEASUREMENT PARAMETERS

Instrumentation is used to measure many parameters (physical values). These parameters include:

• Pressure, either differential or static • Flow • Temperature • Levels of liquids, etc. • Density	• Viscosity • ionising radiation • Frequency • Current	• Voltage • Inductance • Capacitance • Resistivity	• Chemical composition • Chemical properties • Vibration • Weight

Control valve.

INSTRUMENTATION ENGINEERING

Instrumentation engineering is the engineering specialization focused on the principle and operation of measuring instruments that are used in design and configuration of automated systems in electrical, pneumatic domains etc. They typically work for industries with automated processes, such as chemical or manufacturing plants, with the goal of improving system productivity, reliability, safety, optimization, and stability. To control the parameters in a process or in a particular system, devices such as microprocessors, microcontrollers or PLCs are used, but their ultimate aim is to control the parameters of a system.

	T001	P001
SERVICE	STORAGE TANK	FEED PUMP
DATA	DIAMETER: 1000 mm HEIGHT: 3000 mm CAPACITY: 2.4 m3	FLOW RATE: 5 m3/h DIFF. PRESSURE: 2.5 bar
DESIGN PRESSURE	10 barg	10 barg
DESIGN TEMP.	50 °C	50 °C

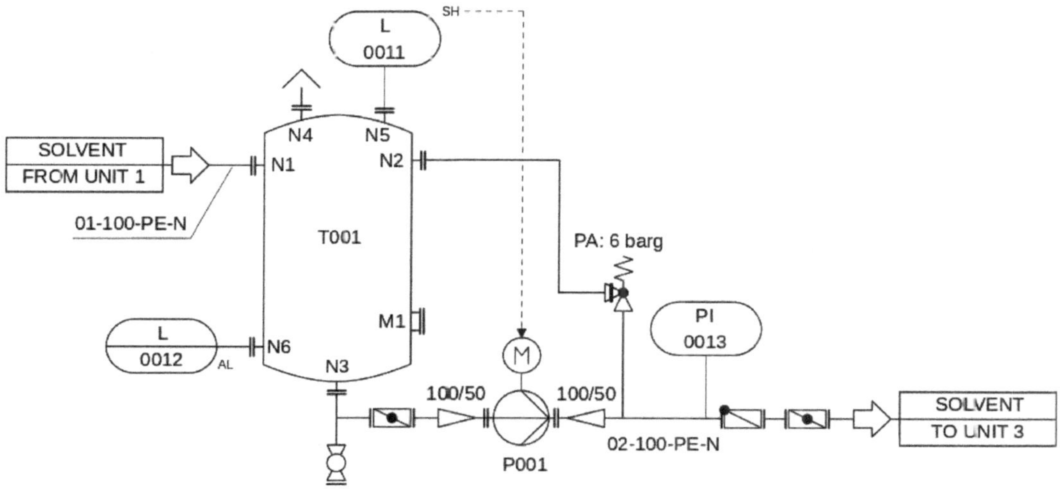

The instrumentation part of a Piping and instrumentation diagram will be developed by an instrumentation engineer.

Instrumentation engineering is loosely defined because the required tasks are very domain dependent. An expert in the biomedical instrumentation of laboratory rats has very different concerns than the expert in rocket instrumentation. Common concerns of both are the selection of appropriate sensors based on size, weight, cost, reliability, accuracy, longevity, environmental robustness and frequency response. Some sensors are literally fired in artillery shells. Others sense thermonuclear explosions until destroyed. Invariably sensor data must be recorded, transmitted or displayed. Recording rates and capacities vary enormously. Transmission can be trivial or can be clandestine, encrypted and low-power in the presence of jamming. Displays can be trivially simple or can require consultation with human factors experts. Control system design varies from trivial to a separate specialty.

Instrumentation engineers are responsible for integrating the sensors with the recorders, transmitters, displays or control systems, and producing the Piping and instrumentation diagram for the process. They may design or specify installation, wiring and signal conditioning. They may be responsible for calibration, testing and maintenance of the system.

In a research environment it is common for subject matter experts to have substantial instrumentation system expertise. An astronomer knows the structure of the universe and a great deal about telescopes - optics, pointing and cameras (or other sensing elements). That often includes the hard-won knowledge of the operational procedures that provide the best results. For example, an astronomer is often knowledgeable of techniques to minimize temperature gradients that cause air turbulence within the telescope.

Instrumentation technologists, technicians and mechanics specialize in troubleshooting, repairing and maintaining instruments and instrumentation systems.

Typical Industrial Transmitter Signal Types

Current Loop (4-20mA) - Electrical

HART - Data signalling often overlaid on a current loop.

Foundation Fieldbus - Data signalling

Profibus - Data signalling

IMPACT OF MODERN DEVELOPMENT

Ralph Müller (1940) stated "That the history of physical science is largely the history of instruments and their intelligent use is well known. The broad generalizations and theories which have arisen from time to time have stood or fallen on the basis of accurate measurement, and in several instances new instruments have had to be devised for the purpose. There is little evidence to show that the mind of modern man is superior to that of the ancients. His tools are incomparably better."

Davis Baird has argued that the major change associated with Floris Cohen's identification of a "fourth big scientific revolution" after World War II is the development of scientific instrumentation, not only in chemistry but across the sciences. In chemistry, the introduction of new instrumentation in the 1940s was "nothing less than a scientific and technological revolution" in which classical wet-and-dry methods of structural organic chemistry were discarded, and new areas of research opened up.

As early as 1954, W A Wildhack discussed both the productive and destructive potential inherent in process control. The ability to make precise, verifiable and reproducible measurements of the natural world, at levels that were not previously observable, using scientific instrumentation, has "provided a different texture of the world". This instrumentation revolution fundamentally changes human abilities to monitor and respond, as is illustrated in the examples of DDT monitoring and the use of UV spectrophotometry and gas chromatography to monitor water pollutants.

REFERENCES

1. "Early Clocks". Retrieved 1 March 2012.
2. "Building automation history page". Retrieved 1 March 2012.

3. Multhauf, Robert P. (1961), The Introduction of Self-Registering Meteorological Instruments, Washington, D.C.: Smithsonian Institution, pp. 95–116 United States National Museum, Bulletin 228. Contributions from The Museum of History and Technology: Paper 23. Available from Project Gutenberg.
4. Anderson, Norman A. (1998). Instrumentation for Process Measurement and Control (3 ed.). CRC Press. pp. 254–255. ISBN 0-8493-9871-1.
5. Anderson, Norman A. (1998). Instrumentation for Process Measurement and Control (3 ed.). CRC Press. pp. 8–10. ISBN 0-8493-9871-1.
6. Aircraft Instrumentation - Leroy R. Grumman Cadet Squadron
7. Katz, Eric; Light, Andrew; Thompson, William (2002). Controlling technology : contemporary issues (2nd ed.). Amherst, NY: Prometheus Books. ISBN 978-1573929837. Retrieved 9 March 2016.
8. Baird, D. (1993). "Analytical chemistry and the 'big' scientific instrumentation revolution". Annals of Science. 50: 267–290. doi:10.1080/00033799300200221. Download the pdf to read the full article.
9. Baird, D. (2002). "Analytical chemistry and the 'big' scientific instrumentation revolution". In Morris, Peter J. T. From classical to modern chemistry : the instrumental revolution ; from a conference on the history of chemical instrumentation: "From the Test-tube to the Autoanalyzer: the Development of Chemical Instrumentation in the Twentieth Century", London, in August 2000. Cambridge: Royal Society of Chemistry in assoc. with the Science Museum. pp. 29–56. ISBN 9780854044795.
10. Reinhardt, Carsten, ed. (2001). Chemical sciences in twentieth century (1st ed.). Weinheim: Wiley-VCH. ISBN 978-3527302710.
11. Wildhack, W. A. (22 October 1954). "Instrumentation--Revolution in Industry, Science, and Warfare". Science. 120 (3121): 15A–15A. doi:10.1126/science.120.3121.15A. Retrieved 9 March 2016.
12. Hentschel, Klaus (2003). "The Instrumental Revolution in Chemistry (Review Essay)". Foundations of Chemistry. 5 (2): 179–183. doi:10.1023/A:1023691917565. Retrieved 8 March 2016.

CHAPTER 9

Microelectromechanical Systems

Chapter Contents

INTRODUCTION

Microelectromechanical systems (MEMS, also written as *micro-electro-mechanical*, *MicroElectroMechanical* or *microelectronic and microelectromechanical systems* and the related *micromechatronics*) is the technology of microscopic devices, particularly those with moving parts. It merges at the nano-scale into nanoelectromechanical systems (NEMS) and nanotechnology. MEMS are also referred to as micromachines in Japan, or *micro systems technology* (*MST*) in Europe.

MEMS are separate and distinct from the hypothetical vision of molecular nanotechnology or molecular electronics. MEMS are made up of components between 1 and 100 micrometres in size (i.e. 0.001 to 0.1 mm), and MEMS devices generally range in size from 20 micrometres to a millimetre (i.e. 0.02 to 1.0 mm). They usually consist of a central unit that processes data (the microprocessor) and several components that interact with the surroundings such as microsensors. At these size scales, the standard constructs of classical physics are not always sufficient. Because of the large surface area to volume ratio of MEMS, surface effects such as electrostatics and wetting dominate over volume effects such as inertia or thermal mass.

The potential of very small machines was appreciated before the technology existed that could make them (see, for example, Richard Feynman's famous 1959 lecture There's Plenty of Room at the Bottom). MEMS became practical once they could be fabricated using modified semiconductor device fabrication technologies, normally used to make electronics. These include molding and plating, wet etching (KOH, TMAH) and dry etching (RIE and DRIE), electro discharge machining (EDM), and other technologies capable of manufacturing small devices. An early example of a MEMS device is the resonistor – an electromechanical monolithic resonator.

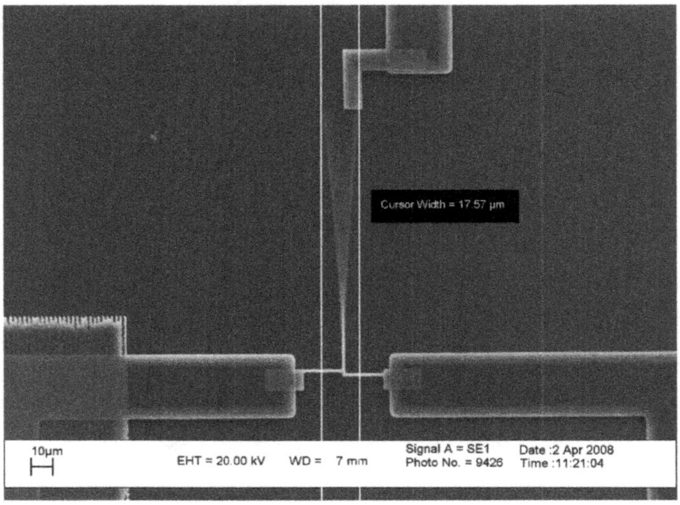

MEMS microcantilever resonating inside a scanning electron microscope.

MATERIALS FOR MEMS MANUFACTURING

The fabrication of MEMS evolved from the process technology in semiconductor device fabrication, i.e. the basic techniques are deposition of material layers, patterning by photolithography and etching to produce the required shapes.

Silicon

Silicon is the material used to create most integrated circuits used in consumer electronics in the modern industry. The economies of scale, ready availability of inexpensive high-quality materials, and ability to incorporate electronic functionality make silicon attractive for a wide variety of MEMS applications. Silicon also has significant advantages engendered through its material properties. In single crystal form, silicon is an almost perfect Hookean material, meaning that when it is flexed there is virtually no hysteresis and hence almost no energy dissipation. As well as making for highly repeatable motion, this also makes silicon very reliable as it suffers very little fatigue and can have service lifetimes in the range of billions to trillions of cycles without breaking.

Polymers

Even though the electronics industry provides an economy of scale for the silicon industry, crystalline silicon is still a complex and relatively expensive material to produce. Polymers on the other hand can be produced in huge volumes, with a great variety of material characteristics. MEMS devices can be made from polymers by processes such as injection molding, embossing or stereolithography and are especially well suited to microfluidic applications such as disposable blood testing cartridges.

Metals

Metals can also be used to create MEMS elements. While metals do not have some of the advantages displayed by silicon in terms of mechanical properties, when used within their limitations, metals can exhibit very high degrees of reliability. Metals can be deposited by electroplating, evaporation, and sputtering processes. Commonly used metals include gold, nickel, aluminium, copper, chromium, titanium, tungsten, platinum, and silver.

Ceramics

The nitrides of silicon, aluminium and titanium as well as silicon carbide and other ceramics are increasingly applied in MEMS fabrication due to advantageous combinations of material properties. AlN crystallizes in the wurtzite structure and thus shows pyroelectric and piezoelectric properties enabling sensors, for instance, with sensitivity to normal and shear forces.TiN, on the other hand, exhibits a high electrical conductivity and large elastic modulus allowing to realize electrostatic MEMS actuation schemes with ultrathin membranes. Moreover, the high resistance of TiN against biocorrosion qualifies the material for applications in biogenic environments and in biosensors.

MEMS BASIC PROCESSES

Deposition Processes

One of the basic building blocks in MEMS processing is the ability to deposit thin films of material with a thickness anywhere between a few nanometres to about 100 micrometres. There are two types of deposition processes, as follows.

Physical Deposition

Physical vapor deposition ("PVD") consists of a process in which a material is removed from a target, and deposited on a surface. Techniques to do this include the process of sputtering, in which an ion beam liberates atoms from a target, allowing them to move through the intervening space and deposit on the desired substrate, and evaporation, in which a material is evaporated from a target using either heat (thermal evaporation) or an electron beam (e-beam evaporation) in a vacuum system.

Chemical Deposition

Chemical deposition techniques include chemical vapor deposition ("CVD"), in which a stream of source gas reacts on the substrate to grow the material desired. This can be further divided into categories depending on the details of the technique, for example, LPCVD (Low Pressure chemical vapor deposition) and PECVD (Plasma Enhanced chemical vapor deposition).

Oxide films can also be grown by the technique of thermal oxidation, in which the (typically silicon) wafer is exposed to oxygen and/or steam, to grow a thin surface layer of silicon dioxide.

Patterning

Patterning in MEMS is the transfer of a pattern into a material.

Lithography

Lithography in MEMS context is typically the transfer of a pattern into a photosensitive material by selective exposure to a radiation source such as light. A photosensitive material is a material that experiences a change in its physical properties when exposed to a radiation source. If a photosensitive material is selectively exposed to radiation (e.g. by masking some of the radiation) the pattern of the radiation on the material is transferred to the material exposed, as the properties of the exposed and unexposed regions differs.

This exposed region can then be removed or treated providing a mask for the underlying substrate. Photolithography is typically used with metal or other thin film deposition, wet and dry etching.

Electron Beam Lithography

Electron beam lithography (often abbreviated as e-beam lithography) is the practice of scanning a beam of electrons in a patterned fashion across a surface covered with a film (called the resist), ("exposing" the resist) and of selectively removing either exposed or non-exposed regions of the resist ("developing"). The purpose, as with photolithography, is to create very small structures in the resist that can subsequently be transferred to the substrate material, often by etching. It was developed for manufacturing integrated circuits, and is also used for creating nanotechnology architectures.

The primary advantage of electron beam lithography is that it is one of the ways to beat the diffraction limit of light and make features in the nanometer region. This form of maskless lithography has found wide usage in photomask-making used in photolithography, low-volume production of semiconductor components, and research & development.

The key limitation of electron beam lithography is throughput, i.e., the very long time it takes to expose an entire silicon wafer or glass substrate. A long exposure time leaves the user vulnerable to beam drift or instability which may occur during the exposure.

Also, the turn-around time for reworking or re-design is lengthened unnecessarily if the pattern is not being changed the second time.

Ion Beam Lithography

It is known that focused-ion beam lithography has the capability of writing extremely fine lines (less than 50 nm line and space has been achieved) without proximity effect. However, because the writing field in ion-beam lithography is quite small, large area patterns must be created by stitching together the small fields.

Ion Track Technology

Ion track technology is a deep cutting tool with a resolution limit around 8 nm applicable to radiation resistant minerals, glasses and polymers. It is capable to generate holes in thin films without any development process. Structural depth can be defined either by ion range or by material thickness. Aspect ratios up to several 10^4 can be reached. The technique can shape and texture materials at a defined inclination angle. Random pattern, single-ion track structures and aimed pattern consisting of individual single tracks can be generated.

X-Ray Lithography

X-ray lithography is a process used in electronic industry to selectively remove parts of a thin film. It uses X-rays to transfer a geometric pattern from a mask to a light-sensitive chemical photoresist, or simply "resist," on the substrate. A series of chemical treatments then engraves the produced pattern into the material underneath the photoresist.

Diamond Patterning

A simple way to carve or create patterns on the surface of nanodiamonds without damaging them could lead to a new photonic devices.

Diamond patterning is a method of forming diamond MEMS. It is achieved by the lithographic application of diamond films to a substrate such as silicon. The patterns can be formed by selective deposition through a silicon dioxide mask, or by deposition followed by micromachining or focused ion beam milling.

Etching Processes

There are two basic categories of etching processes: wet etching and dry etching. In the former, the material is dissolved when immersed in a chemical solution. In the latter, the material is sputtered or dissolved using reactive ions or a vapor phase etchant.

Wet Etching

Wet chemical etching consists in selective removal of material by dipping a substrate into a solution that dissolves it. The chemical nature of this etching process provides a good selectivity, which means the etching rate of the target material is considerably higher than the mask material if selected carefully.

Isotropic Etching

Etching progresses at the same speed in all directions. Long and narrow holes in a mask will produce v-shaped grooves in the silicon. The surface of these grooves can be atomically smooth if the etch is carried out correctly, with dimensions and angles being extremely accurate.

Anisotropic Etching

Some single crystal materials, such as silicon, will have different etching rates depending on the crystallographic orientation of the substrate. This is known as anisotropic etching and one of the most common examples is the etching of silicon in KOH (potassium hydroxide), where Si <111> planes etch approximately 100 times slower than other planes (crystallographic orientations). Therefore, etching a rectangular hole in a (100)-Si wafer results in a pyramid shaped etch pit with 54.7° walls, instead of a hole with curved sidewalls as with isotropic etching.

HF Etching

Hydrofluoric acid is commonly used as an aqueous etchant for silicon dioxide (SiO2, also known as BOX for SOI), usually in 49% concentrated form, 5:1, 10:1 or 20:1 BOE (buffered oxide etchant) or BHF (Buffered HF). They were first used in medieval times for glass etching. It was used in IC fabrication for patterning the gate oxide until the process step was replaced by RIE.

Hydrofluoric acid is considered one of the more dangerous acids in the cleanroom. It penetrates the skin upon contact and it diffuses straight to the bone. Therefore, the damage is not felt until it is too late.

Electrochemical Etching

Electrochemical etching (ECE) for dopant-selective removal of silicon is a common method to automate and to selectively control etching. An active p-n diode junction is required, and either type of dopant can be the etch-resistant ("etch-stop") material. Boron is the most common etch-stop dopant. In combination with wet anisotropic etching as described above, ECE has been used successfully for controlling silicon diaphragm thickness in commercial piezoresistive silicon pressure sensors. Selectively doped regions can be created either by implantation, diffusion, or epitaxial deposition of silicon.

Dry Etching

Vapor Etching

Xenon Difluoride

Xenon difluoride (XeF_2) is a dry vapor phase isotropic etch for silicon originally applied for MEMS in 1995 at University of California, Los Angeles. Primarily used for releasing metal and dielectric structures by undercutting silicon, XeF_2 has the advantage of a stiction-free release unlike wet etchants. Its etch selectivity to silicon is very high, allowing it to work with photoresist, SiO_2, silicon nitride, and various metals for masking. Its reaction to silicon is "plasmaless", is purely chemical and spontaneous and is often operated in pulsed mode. Models of the etching action are available, and university laboratories and various commercial tools offer solutions using this approach.

Plasma Etching

Modern VLSI processes avoid wet etching, and use plasma etching instead. Plasma etchers can operate in several modes by adjusting the parameters of the plasma. Ordinary plasma etching operates between 0.1 and 5 Torr. (This unit of pressure, commonly used in vacuum engineering, equals approximately 133.3 pascals.) The plasma produces energetic free radicals, neutrally charged, that react at the surface of the wafer. Since neutral particles attack the wafer from all angles, this process is isotropic.

Plasma etching can be isotropic, i.e., exhibiting a lateral undercut rate on a patterned surface approximately the same as its downward etch rate, or can be anisotropic, i.e., exhibiting a smaller lateral undercut rate than its downward etch rate. Such anisotropy is maximized in deep reactive ion etching. The use of the term anisotropy for plasma etching should not be conflated with the use of the same term when referring to orientation-dependent etching.

The source gas for the plasma usually contains small molecules rich in chlorine or fluorine. For instance, carbon tetrachloride (CCl_4) etches silicon and aluminium, and trifluoromethane etches silicon dioxide and silicon nitride. A plasma containing oxygen is used to oxidize ("ash") photoresist and facilitate its removal.

Ion milling, or sputter etching, uses lower pressures, often as low as 10^{-4} Torr (10 mPa). It bombards the wafer with energetic ions of noble gases, often Ar^+, which knock atoms from the substrate by transferring momentum. Because the etching is performed by ions, which approach the wafer approximately from one direction, this process is highly anisotropic. On the other hand, it tends to display poor selectivity. Reactive-ion etching (RIE) operates under conditions intermediate between sputter and plasma etching (between 10^{-3} and 10^{-1} Torr). Deep reactive-ion etching (DRIE) modifies the RIE technique to produce deep, narrow features.

Sputtering

Reactive ion Etching (RIE)

In reactive-ion etching (RIE), the substrate is placed inside a reactor, and several gases are introduced. A plasma is struck in the gas mixture using an RF power source, which breaks the gas molecules into ions. The ions accelerate towards, and react with, the surface of the material being etched, forming another gaseous material. This is known as the chemical part of reactive ion etching. There is also a physical part, which is similar to the sputtering deposition process. If the ions have high enough energy, they can knock atoms out of the material to be etched without a chemical reaction. It is a very complex task to develop dry etch processes that balance chemical and physical etching, since there are many parameters to adjust. By changing the balance it is possible to influence the anisotropy of the etching, since the chemical part is isotropic and the physical part highly anisotropic the combination can form sidewalls that have shapes from rounded to vertical.

Deep reactive ion etching

Deep RIE (DRIE) is a special subclass of RIE that is growing in popularity. In this process, etch depths of hundreds of micrometres are achieved with almost vertical sidewalls. The primary technology is based on the so-called "Bosch process", named after the German company Robert Bosch, which filed the original patent, where two different gas compositions alternate in the reactor. Currently there are two variations of the DRIE. The first variation consists of three distinct steps (the original Bosch process) while the second variation only consists of two steps.In the first variation, the etch cycle is as follows:(i) SF6 isotropic etch;(ii) C4F8 passivation;(iii) SF6 anisoptropic etch for floor cleaning.In the 2nd variation, steps (i) and (iii) are combined.

Both variations operate similarly. The C4F8 creates a polymer on the surface of the substrate, and the second gas composition (SF6 and O2) etches the substrate. The polymer is immediately sputtered away by the physical part of the etching, but only on the horizontal surfaces and not the sidewalls. Since the polymer only dissolves very slowly in the chemical part of the etching, it builds up on the sidewalls and protects them from etching. As a result, etching aspect ratios of 50 to 1 can be achieved. The process can easily be used to etch completely through a silicon substrate, and etch rates are 3–6 times higher than wet etching.

Die Preparation

After preparing a large number of MEMS devices on a silicon wafer, individual dies have to be separated, which is called die preparation in semiconductor technology. For some applications, the separation is preceded by wafer backgrinding in order to reduce the wafer thickness. Wafer dicing may then be performed either by sawing using a cooling liquid or a dry laser process called stealth dicing.

MEMS MANUFACTURING TECHNOLOGIES

Bulk Micromachining

Bulk micromachining is the oldest paradigm of silicon based MEMS. The whole thickness of a silicon wafer is used for building the micro-mechanical structures. Silicon is machined using various etching processes. Anodic bonding of glass plates or additional silicon wafers is used for adding features in the third dimension and for hermetic encapsulation. Bulk micromachining has been essential in enabling high performance pressure sensors and accelerometers that changed the sensor industry in the 1980s and 90's.

Surface Micromachining

Surface micromachining uses layers deposited on the surface of a substrate as the structural materials, rather than using the substrate itself. Surface micromachining was created in the late 1980s to render micromachining of silicon more compatible with planar integrated circuit technology, with the goal of combining MEMS and integrated circuits on the same silicon wafer. The original surface micromachining concept was based on thin polycrystalline silicon layers patterned as movable mechanical structures and released by sacrificial etching of the underlying oxide layer. Interdigital comb electrodes were used to produce in-plane forces and to detect in-plane movement capacitively. This MEMS paradigm has enabled the manufacturing of low cost accelerometers for e.g. automotive air-bag systems and other applications where low performance and/or high g-ranges are sufficient. Analog Devices has pioneered the industrialization of surface micromachining and has realized the co-integration of MEMS and integrated circuits.

High Aspect Ratio (HAR) Silicon Micromachining

Both bulk and surface silicon micromachining are used in the industrial production of sensors, ink-jet nozzles, and other devices. But in many cases the distinction between these two has diminished. A new etching technology, deep reactive-ion etching, has made it possible to combine good performance typical of bulk micromachining with comb structures and in-plane operation typical of surface micromachining. While it is common in surface micromachining to have structural layer thickness in the range of 2 μm, in HAR silicon micromachining the thickness can be from 10 to 100 μm. The materials commonly used in HAR silicon micromachining are thick polycrystalline silicon, known as epi-poly, and bonded silicon-on-insulator (SOI) wafers although processes for bulk silicon wafer also have been created (SCREAM). Bonding a second wafer by glass frit bonding, anodic bonding or alloy bonding is used to protect the MEMS structures. Integrated circuits are typically not combined with HAR silicon micromachining.

Microelectromechanical systems chip, sometimes called "lab on a chip"

APPLICATIONS

A Texas Instruments DMD chip for cinema projection

Some common commercial applications of MEMS include:

- Inkjet printers, which use piezoelectrics or thermal bubble ejection to deposit ink on paper.
- Accelerometers in modern cars for a large number of purposes including airbag deployment and electronic stability control.

- Accelerometers and MEMS gyroscopes in remote controlled, or autonomous, helicopters, planes and multirotors (also known as drones), used for automatically sensing and balancing flying characteristics of roll, pitch and yaw.
- Accelerometers in consumer electronics devices such as game controllers (Nintendo Wii), personal media players / cell phones (Apple iPhone, various Nokia mobile phone models, various HTC PDA models) and a number of Digital Cameras (various Canon Digital IXUS models). Also used in PCs to park the hard disk head when free-fall is detected, to prevent damage and data loss.
- MEMS gyroscopes used in modern cars and other applications to detect yaw; e.g., to deploy a roll over bar or trigger electronic stability control
- MEMS microphones in portable devices, e.g., mobile phones, head sets and laptops. The market for smart microphones includes smartphones, wearable devices, smart home and automotive applications.
- Silicon pressure sensors e.g., car tire pressure sensors, and disposable blood pressure sensors
- Displays e.g., the digital micromirror device (DMD) chip in a projector based on DLP technology, which has a surface with several hundred thousand micromirrors or single micro-scanning-mirrors also called microscanners
- Optical switching technology, which is used for switching technology and alignment for data communications
- Bio-MEMS applications in medical and health related technologies from Lab-On-Chip to MicroTotalAnalysis (biosensor, chemosensor), or embedded in medical devices e.g. stents.
- Interferometric modulator display (IMOD) applications in consumer electronics (primarily displays for mobile devices), used to create interferometric modulation – reflective display technology as found in mirasol displays
- Fluid acceleration such as for micro-cooling
- Micro-scale energy harvesting including piezoelectric, electrostatic and electromagnetic micro harvesters.
- Micromachined ultrasound transducers.

INDUSTRY STRUCTURE

The global market for micro-electromechanical systems, which includes products such as automobile airbag systems, display systems and inkjet cartridges totaled $40 billion in 2006 according to Global MEMS/Microsystems Markets and Opportunities, a research report from SEMI and Yole Developpement and is forecasted to reach $72 billion by 2011.

Companies with strong MEMS programs come in many sizes. Larger firms specialize in manufacturing high volume inexpensive components or packaged solutions for end markets such as automobiles, biomedical, and electronics. Smaller firms provide value in innovative solutions and absorb the expense of custom fabrication with high sales margins. Both large and small companies typically invest in R&D to explore new MEMS technology.

The market for materials and equipment used to manufacture MEMS devices topped $1 billion worldwide in 2006. Materials demand is driven by substrates, making up over 70 percent of the market, packaging coatings and increasing use of chemical mechanical planarization (CMP). While MEMS manufacturing continues to be dominated by used semiconductor equipment, there is a migration to 200 mm lines and select new tools, including etch and bonding for certain MEMS applications.

REFERENCES

1. Waldner, Jean-Baptiste (2008). Nanocomputers and Swarm Intelligence. London: ISTE John Wiley & Sons. p. 205. ISBN 1-84821-009-4.
2. James B. Angell; Stephen C. Terry; Phillip W. Barth (April 1983). "Silicon Micromechanical Devices". Scientific American. 248 (4): 44–55.
3. Electromechanical monolithic resonator, US patent 3614677, Filed April 29, 1966; Issued October 1971
4. Wilfinger, R.J.; Bardell, P.H.; Chhabra, D.S. (1968). "The Resonistor: A Frequency Selective Device Utilizing the Mechanical Resonance of a Silicon Substrate". IBM J. 12: 113–8. doi:10.1147/rd.121.0113.
5. R. Ghodssi; P. Lin (2011). MEMS Materials and Processes Handbook. Berlin: Springer. ISBN 978-0-387-47316-1.
6. T. Polster; M. Hoffmann (2009). "Aluminium nitride based 3D, piezoelectric, tactile sensors". Proc. Chem. 1: 144–7. doi:10.1016/j.proche.2009.07.036.
7. M. Birkholz; K.-E. Ehwald; P. Kulse; J. Drews; M. Fröhlich; U. Haak; M. Kaynak; E. Matthus; K. Schulz; D. Wolansky (2011). "Ultrathin TiN membranes as a technology platform for CMOS-integrated MEMS and BioMEMS devices" (PDF). Adv. Func. Mat. 21 (9): 1652–1654. doi:10.1002/adfm.201002062.
8. McCord, M. A.; M. J. Rooks (2000). "2". SPIE Handbook of Microlithography, Micromachining and Microfabrication.
9. Marc J. Madou, Fundamentals of Microfabrication and Nanotechnology, Volume III: From MEMS to Bio-MEMS and Bio-NEMS: Manufacturing Techniques and Applications, p. 252, CRC Press, 2011 ISBN 1439895244.
10. Williams, K.R.; Muller, R.S. (1996). "Etch rates for micromachining processing". Journal of Microelectromechanical Systems. 5 (4): 256. doi:10.1109/84.546406.
11. Kovacs, G.T.A.; Maluf, N.I.; Petersen, K.E. (1998). "Bulk micromachining of silicon". Proceedings of the IEEE. 86 (8): 1536. doi:10.1109/5.704259.
12. Chang, Floy I. (1995). "Gas-phase silicon micromachining with xenon difluoride". Gas-phase silicon micromachining with xenon difluoride. Microelectronic Structures and Microelectromechanical Devices for Optical Processing and Multimedia Applications. 2641. p. 117. doi:10.1117/12.220933.
13. Chang, Floy I-Jung (1995). Xenon difluoride etching of silicon for MEMS (M.S.). Los Angeles: University of California. OCLC 34531873.

14. Brazzle, J.D.; Dokmeci, M.R.; Mastrangelo, C.H. (2004). "Modeling and characterization of sacrificial polysilicon etching using vapor-phase xenon difluoride". 17th IEEE International Conference on Micro Electro Mechanical Systems. Maastricht MEMS 2004 Technical Digest. p. 737. doi:10.1109/MEMS.2004.1290690. ISBN 0-7803-8265-X.

15. Laermer, F.; Urban, A. (2005). "Milestones in deep reactive ion etching". The 13th International Conference on Solid-State Sensors, Actuators and Microsystems, 2005. Digest of Technical Papers. TRANSDUCERS '05. 2. p. 1118. doi:10.1109/SENSOR.2005.1497272. ISBN 0-7803-8994-8.

16. Bustillo, J. M.; Howe, R. T.; Muller, R. S. (August 1998). "Surface Micromachining for Microelectromechanical Systems" (PDF). Proceedings of the IEEE. 86 (8): 1552–1574. CiteSeerX 10.1.1.120.4059Freely accessible. doi:10.1109/5.704260.

17. Hosseinian, Ehsan; Pierron, Olivier N. (2013). "Quantitative in situ TEM tensile fatigue testing on nanocrystalline metallic ultrathin films". Nanoscale. 5 (24): 12532. doi:10.1039/C3NR04035F. PMID 24173603.

18. Johnson, R. Collin (2007-07-09). There's more to MEMS than meets the iPhone, EE Times

19. Cenk Acar; Andrei M. Shkel (2008). MEMS Vibratory Gyroscopes: Structural Approaches to Improve Robustness. pp. 111 ff. ISBN 0-387-09535-7.

20. By Peter Clarke, EE Times Europe. "Smart MEMS microphones market emerges." May 31, 2016. Retrieved June 1, 2016.

21. Louizos, Louizos-Alexandros; Athanasopoulos, Panagiotis G.; Varty, Kevin (2012). "Microelectromechanical Systems and Nanotechnology. A Platform for the Next Stent Technological Era". Vasc Endovascular Surg. 46 (8): 605–609. doi:10.1177/1538574412462637. PMID 23047818.

22. Hajati, Arman; Sang-Gook Kim (2011). "Ultra-wide bandwidth piezoelectric energy harvesting". Applied Physics Letters. 99 (8): 083105. doi:10.1063/1.3629551.

23. Hajati, Arman (2012). "Three-dimensional micro electromechanical system piezoelectric ultrasound transducer". Applied Physics Letters. 101 (25): 253101. doi:10.1063/1.4772469.

24. Hajati, Arman (2013). "Monolithic ultrasonic integrated circuits based on micromachined semi-ellipsoidal piezoelectric domes". Applied Physics Letters. 103 (20): 202906. doi:10.1063/1.4831988.

25. Worldwide MEMS Systems Market Forecasted to Reach $72 Billion by 2011. Azonano.com (2007-07-17). Retrieved on 2015-10-05.

Index